Universal and Accessible Design for Products, Services, and Processes

Universal and Accessible Design for Products, Services, and Processes

Robert F. Erlandson

CRC Press
Taylor & Francis Group
Boca Raton London New York

CRC Press is an imprint of the
Taylor & Francis Group, an **informa** business

CRC Press
Taylor & Francis Group
6000 Broken Sound Parkway NW, Suite 300
Boca Raton, FL 33487-2742

© 2008 by Taylor & Francis Group, LLC
CRC Press is an imprint of Taylor & Francis Group, an Informa business

No claim to original U.S. Government works
Printed in the United States of America on acid-free paper
10 9 8 7 6 5 4 3 2 1

International Standard Book Number-13: 978-0-8493-7493-7 (Hardcover)

This book contains information obtained from authentic and highly regarded sources. Reprinted material is quoted with permission, and sources are indicated. A wide variety of references are listed. Reasonable efforts have been made to publish reliable data and information, but the author and the publisher cannot assume responsibility for the validity of all materials or for the consequences of their use.

No part of this book may be reprinted, reproduced, transmitted, or utilized in any form by any electronic, mechanical, or other means, now known or hereafter invented, including photocopying, microfilming, and recording, or in any information storage or retrieval system, without written permission from the publishers.

For permission to photocopy or use material electronically from this work, please access www. copyright.com (http://www.copyright.com/) or contact the Copyright Clearance Center, Inc. (CCC) 222 Rosewood Drive, Danvers, MA 01923, 978-750-8400. CCC is a not-for-profit organization that provides licenses and registration for a variety of users. For organizations that have been granted a photocopy license by the CCC, a separate system of payment has been arranged.

Trademark Notice: Product or corporate names may be trademarks or registered trademarks, and are used only for identification and explanation without intent to infringe.

Library of Congress Cataloging-in-Publication Data

Erlandson, Robert F.
    Universal and accessible design for products, services, and processes / Robert F. Erlandson.
         p. cm.
    Includes bibliographical references and index.
    ISBN-13: 978-0-8493-7493-7 (alk. paper)
    ISBN-10: 0-8493-7493-6 (alk. paper)
    1. Barrier-free design. I. Title.

NA2545.A1E75 2008
725'.54--dc22   2007011402

Visit the Taylor & Francis Web site at
http://www.taylorandfrancis.com

and the CRC Press Web site at
http://www.crcpress.com

Dedication

This book is dedicated to the memory of my parents

Ivar Erlandson (November 21, 1905 to March 8, 2006)

and

Mildred Erlandson (December 25, 1907 to July 24, 2006)

Accessible design 20
Adaptable design and accessible design 20
Adaptable, accessible, and universal design 21
Adaptable design 21
Universal design in a global market 22
Conclusion 23
References 24

Section 2: Disability, laws, and accessible design

Section goals 27
Scope of the discussion 27

Chapter 3 Disability and design 29
Chapter goals 29
Conceptual models 29
Definition of disability 30
Enabling and disabling environmental factors 32
Design implications associated with the WHO-ICF conceptual model 32
    Medical model 33
    Social model 33
    Systems model 34
Accessible design, universal design, and conceptual models
    of disability 34
Conceptual model for this book 35
References 35

Chapter 4 Accessible design 37
Chapter goals 37
Access Board 37
Laws and accessibility 42
    Architectural Barriers Act (ABA) of 1968 as amended
        42 U.S.C. § 4151 et seq. 42
    1990 Americans with Disabilities Act (ADA) 42
        Title I 42
        Titles II and III 44
        ADA and Web accessibility 46
        Title IV 46
    Telecommunication: Section 255, Telecommunications Act of 1996 46
    Compatibility 49
    Electronic and information technology: Section 508 of the
        Rehabilitation Act Amendments of 1998 51
Access Board accessible design technical assistance 54

Accessibility forum 54
Conclusions 58
References 59

Section 3: Universal design principles, strategies, and examples

Section goals 63
Scope of the discussion 63

Chapter 5 Hierarchical structure of universal design principles and implications for the design process 67
Chapter goals 67
Design philosophy 67
Universal design principles 67
Principles targeted at the person (human factors) 68
Principles targeted at the process 69
Transcending, integrating principle 69
Hierarchical constraints 70
References 71

Chapter 6 Ergonomically sound 73
Chapter goals 73
Ergonomically sound 73
    Principle 73
    Discussion 73
Design strategies 76
    Strategy 1: Design to avoid ergonomic risk factors 76
    Strategy 2: Design for a wide range of body sizes and shapes 77
    Strategy 3: Design for ease of use 80
Conclusions 81
References 81

Chapter 7 Perceptible 83
Chapter goals 83
Perceptible 83
    Principle 83
    Discussion 83
Design strategies 84
    Strategy 4: Provide multisensory options for communications
        between a person and the process or product 84
    Strategy 5: Design signals so as to maximize the
        signal-to-noise ratio 85
        Signal detection 86

Weber's law 86
Fechner's law 87
Stevens' power law 87
Strategy 5.1: Provide the ability for a person to increase or decrease the signal strength so as to increase the signal-to-noise ratio 88
Strategy 5.2: Provide sufficient contrast between signals and background ambient conditions so that most people will be able to perceive the signal and the message conveyed by the signal 89
Judgments 90
Strategy 6: Keep the signaling structure and content as simple as possible 91
Movement: Reaching, picking, and placing 92
Strategy 7: Design entities so that a user can accurately acquire a target 92
Conclusions 94
References 94

Chapter 8 Cognitively sound 95
Chapter goals 95
Cognitively sound 95
Principle 95
Discussion 95
Design strategies 97
Strategy 8: Build knowledge into the designed entity or environment 97
Affordance 97
Strategy 8.1: Use affordances to help users form clear conceptual models of the entity's operations 97
Mappings and representation 99
Strategy 8.2: Use mappings to help users form clear conceptual models of the entity's operations and simplify operations 99
Constraints 103
Strategy 8.3: Use constraints so as to control the course of actions and prevent or reduce the possibility of the users doing the wrong thing 103
Feedback 103
Strategy 8.4: Use feedback to keep the user informed as to the status of the entity's operations and the entity's response to user inputs 103
Language 104

Contents

    Strategy 9: Provide messages in a language and format that the people using the process or product will understand 105
        Strategy 9.1: Use a language that users understand (English, Spanish, etc.) 105
        Strategy 9.2: Use an appropriate style of language—formal/informal, technical/nontechnical 105
        Strategy 9.3: Use an appropriate level (grade level) of language 105
        Strategy 9.4: Use universally or globally understood icons, symbols, or pictures for communications 106
Complexity 106
    Strategy 10: Reduce the operational complexity of the entity 106
        Strategy 10.1: Keep the task and activity structure as simple as possible 107
        Strategy 10.2: Design an entity so that the user's reaction time is within satisfactory bounds 108
        Strategy 10.3: Keep memory requirements simple 110
        Strategy 10.4: Avoid the use of modes 111
        Education and training 114
    Strategy 11: Use universal design for learning strategies developed by CAST for training or educational activities 114
        Strategy 11.1: Provide the user with multiple means of engagement 114
        Strategy 11.2: Provide the user with multiple means of representation 115
        Strategy 11.3: Provide the user with multiple means of expression 115
Conclusions 115
References 116

Chapter 9   Flexible 119
Chapter goals 119
Flexible 119
    Principle 119
    Discussion 119
    Strategy 12: Provide the user with choices 120
        Strategy 12.1: Provide the user with a choice of language (English, French, etc.) 121
        Strategy 12.2: Provide the user with a choice of mode for communication 122

Strategy 12.3: Ensure that appropriate assistive
   technologies can be effectively used with the
   designed entity (that is, ensure compatibility
   with appropriate assistive technology)......................122
Strategy 13: Provide adjustability and mobility 123
   Strategy 13.1: Provide ergonomic and environmental
      adjustability 123
   Strategy 13.2: Provide perceptual adjustability.......................124
   Strategy 13.3: Provide adjustable response times...................124
   Strategy 13.4: Use agile system design strategies—
      adjustable and mobile..125
Strategy 14: Build flexibility into service delivery systems and
   work processes...126
   Strategy 14.1: Provide the user with choices for
      educational or training activities consistent
      with the CAST universal design for learning
      strategies presented in strategy 11..............................127
A design challenge 128
Conclusions..128
References..129

Chapter 10   Error-managed (proofed)..131
Chapter goals..131
Error-managed (error-proofed)..131
   Principle..131
   Discussion 131
Design strategies 133
   Slips..133
   Strategy 15: Use a three-staged approach to error-proofing 133
      Strategy 15.1: Prevent errors at the source 133
      Strategy 15.2: Provide a warning that an error has
         or is about to occur 135
      Strategy 15.3: Provide quick and easy recovery if
         an error has occurred..138
   Mistakes..139
Error-proofing for training and educational processes.............140
Conclusions...143
References..144

Chapter 11   Efficient (muda elimination) 147
Chapter goals..147
Efficient (muda elimination) 147
   Principle..147
   Discussion 147

Contents   xiii

Design strategies 149
    Strategy 16: Reduce or eliminate non-value-added
        activity (NVAA) 149
        Strategy 16.1: Use quality control and reliability
            engineering techniques to reduce or eliminate
            NVAA due to repair and rejects, and excessive
            waiting due to setup and equipment breakdown 149
        Strategy 16.2: Make the designed entity as simple and
            easy to use as possible 150
    Appliances versus general-purpose devices 150
    Process and service considerations 152
        Strategy 16.3: Avoid complexity in that it leads to NVAA 152
        Strategy 16.4: Use task analysis techniques to identify
            tasks or activities that can be eliminated or
            redesigned so as to reduce or eliminate NVAA 153
Conclusions 154
References 155

Chapter 12 Stable and predictable 157
Chapter goals 157
Stable and predictable 157
    Principle 157
    Discussion 157
Design strategies 159
    Strategy 17: Work to establish national and international
        standards for products, processes, and services so as to
        reduce their common cause variability 159
    Strategy 18: Reduce the common cause variability
        associated with the person's interaction with the
        product or process 161
    Strategy 19: Reduce common cause variability using quality
        control and reliability engineering techniques to ensure
        proper functioning of the product 162
    Strategy 20: Reduce common cause variability associated with
        process use, this includes work-related processes as
        well as service operations 163
        Strategy 20.1: Reduce common cause variability using
            quality control and reliability engineering
            techniques to ensure proper functioning of
            the process 163
        Strategy 20.2: Reduce the common cause variability
            associated with environmental factors and
            process control and management practices 166
        Strategy 20.2.1: Utilize the five Ss 167

Strategy 20.2.2: Utilize workplace organization 168
Strategy 20.2.3: Utilize standardized
         work procedures 169
Strategy 20.2.4: Utilize visual controls 169
Examples of the strategies and synergies 170
   Example 1: Meter reading 170
   Example 2: Airport 172
Conclusions 174
References 175

Chapter 13   Equitable: Transcending, integrating 179
Chapter goals 179
Equitable: Transcending, integrating 179
   Principle 179
   Discussion 179
Design strategies 180
   Strategy 21: Design entities that are age and context appropriate 180
   Strategy 22: Design entities that are aesthetically pleasing 180
   Strategy 23: Design entities that are competitively priced 182
   Strategy 24: Market the entity for as broad a demographic and
         socioeconomic base as reasonable and possible 184
Transcending and integrating 185
References 185

Chapter 14   Universal and accessible design in the workplace 187
Chapter goals 187
Consumer applications versus workplace applications 187
Implementation of universal design in the workplace: Examples 188
   Fuji Heavy Industries Ltd. 188
   Retail and fast food industries 190
      Retail: Wal-Mart 190
      Fast Food: McDonald's 191
      Comparisons: Wal-Mart and McDonald's 192
   A small company: Art for a Cause 192
   Small retail operations: Goodwill Industries 193
Nature of the industry, size of the company, and scale of operations 196
Design tools for work 197
Conclusions 197
References 198

Chapter 15   The World Wide Web: Accessibility and
         universal design 201
Chapter goals 201

Contents xv

Introduction 201
Infrastructure 202
The World Wide Web Consortium (W3C) and the Web
     Accessibility Initiative (WAI) 203
Design principles of the Web 203
Relationship of the Web content accessibility guidelines to
     accessible and universal design principles 204
Section 508 requirements 207
Emerging products, standards, guidelines, and Web design
     recommendations 211
Authoring tools 213
United States government resources 214
Additional resources 215
Conclusions 215
References 216

Section 4: Ethical considerations and conclusions

Section goals 219
Scope of the discussion 219

Chapter 16   Ethics and universal design 221
Chapter goals 221
Ethics 221
     Ethics of disability 221
Native endowments, the environment, and disability 224
Universal design: Expanding the scope 226
Conclusions 229
References 229

Chapter 17   Universal and accessible design from a social and
     political perspective 231
Chapter goals 231
Linkages 231
From here to where? 232
Universal design from a systems perspective 234
Examples of linkage complexity and the changing role of designers 235
Universal and accessible design in relation to society and politics 237
Conclusions 241
References 242

Index 245

Acknowledgments

Two National Science Foundation (NSF) grants and other public and private sponsors provided support for the creation of this book. The NSF grants were "Proof-of-Concept: Accessible Design Curriculum and Educational Materials" (DUE-9972403) and "Accessible Design Curriculum and Educational Materials" (DUE-0088807). Any opinions, findings, conclusions, or recommendations expressed in the material are those of the author and do not necessarily reflect the views of any of the sponsors.

I want to acknowledge the support of David Sant, ETL Engineer, for countless discussions and collaboration on the design and implementation of the ETL devices and services mentioned in the book; Ellen Traschenko for her help in editing; Dr. Mumtaz Usmen for his critical review of the first draft; and the many Wayne State University students who listened to lectures and participated in class exercises as the material was taking shape into an organized body of knowledge.

Introduction

Goal of this book

The goal of this book is to present universal and accessible design to the U.S. design community. Universal design and accessible design are often used interchangeably, and this can be confusing. While universal design and accessible design share a common core of design principles and strategies, they differ in that accessible design is legally mandated whereas universal design is not. The legal mandates and design guidelines associated with accessible design require that designs be compliant subject to legal penalties. Universal design, on the other hand, is motivated more by global competition and the creation of more universally accessible and usable products and services demanded by international markets.

After reading this book, designers will be able to apply universal and accessible design principles to their specific projects, secure in the knowledge of how their applications fit into a broader societal and competitive global design environment.

Who should read this book and why

Professional designers, educators in engineering and the design professions, and students of design should read this book. Growing pressures for the mainstreaming of universal and accessible design, that is, the integration of the ideas, concepts, strategies, and methods of universal and accessible design into the general knowledge and practice of the design community include laws, global markets and competition, an aging population, technological advances, and evolving conceptualizations of disability and societal values. These pressures provide different, but complementary, reasons for reading this book.

A growing collection of national and international laws is mandating accessible design. Lack of compliance can result in fines, penalties, legal fees, and the costs associated with redesign, retooling for manufacturing, or modifying existing structures. Professionals involved in design activities need to know what laws apply to their specific situations and then where to go to get design guidelines, regulations, and compliance information. Educators dealing with design programs need to understand accessible design and its role within society so as to better prepare their students for professional carriers in design. Graduates from these design programs need to be prepared for today's global job market, and knowledge of accessible design principles is emerging as a core component of that professional preparation.

While compliance is important, it is critical for designers to move past simple compliance and the reactive mind-set it can foster. Universal design provides a broader, yet complementary, approach to design. Universal design has evolved from ethical and market-driven pressures, particularly the pressures of global competition. The Council of Europe has passed a resolution seeking to have universal design principles and methods incorporated into all training and educational programs dealing with design. Toyota has embraced the use of universal design principles throughout its corporate activities. The benefits of universal design in today's highly competitive, global, and e-driven economy will be presented. It is critical for the U.S. design community to understand what universal design has to offer and why U.S. business competitors around the world are moving to adopt and utilize universal design principles.

Organization of the book

This book is divided into four major sections. Section I provides the societal and global context for accessible and universal design. It also provides definitions and examples of accessible and universal design. Section II focuses on accessible design, the legal mandates, guidelines, and U.S. government regulatory and resource agencies. Section III presents universal design principles, strategies, examples, and applications. Section IV explores the expanding conceptualizations of universal design and discusses future directions.

Section I contains two chapters. Chapter 1 paints the "big picture" within which universal and accessible design is growing and evolving. It provides a broad-brush overview of the societal and global issues that have spawned and continue to nurture the growth of accessible and universal design. These issues include the laws mandating accessibility and global market pressures on not only multinational corporations, but also small business. Chapter 2 defines universal and accessible design, and explores their relationship to the broader concept of design. Chapter 2 also presents a variety of examples that exemplify and clarify the definitions and relationships among the design approaches.

The two chapters in Section II focus on legal issues and the conceptualization of accessible design that derives from a legal base. Society's conception of disability is reflected in a country's laws, and a variety of laws currently mandate accessible design. Chapter 3 looks at disability and the relationships between various conceptual models of disability and how these conceptual models have evolved and influenced our laws and our perceptions of accessible and universal design. It is impossible to understand accessible design without considering the societal reason for its existence. Accessible design is driven by the laws mandating the accessibility of buildings, services, telecommunication, and electronic and information technologies for people with disabilities. Society's understanding of disability lies at the heart of these laws. The World Health Organization's new International Classification of Functioning, Disability, and Health (WHO ICF) will serve as the conceptual model for disability for this book. The ICF framework explicitly includes environmental factors, the societal elements so strongly linked to design processes, and thereby the ICF serves as an integrating agent for the concepts presented.

Chapter 4 covers accessible design, the accessibility laws, and resources created to help designers comply with legal mandates. This is the "how-to" part of the book with respect to accessible design. This includes a discussion of the duties and responsibilities of the Access Board, which was created by the laws that mandate accessibility. The legally prescribed mandates and design guidelines developed and provided by the Access Board are an explicit formulation of accessible design strategies and requirements. Designers in the industries covered by the accessibility laws must apply these accessible design strategies to create accessible products and services. The Access Board is also charged with the task of educating designers about the accessibility design guidelines and legal mandates. As such, the Access Board provides a rich source of educational and training material. Chapter 4 contains information that is essential for designers working in areas covered by the accessibility laws.

Section III, "Universal Design Principles, Strategies, and Examples," contains 11 chapters and is the heart of the book with respect to the presentation of universal design principles, strategies, and applications. This is the "how-to" part of the book with respect to universal design. For each universal design principle, a collection of design strategies is presented. The design strategies are followed by examples and applications. The term "design strategy" means concrete, pragmatic ways to implement and build entities so that they, in fact, possess the properties and characteristics espoused by the universal design principles. The design strategies derive from a variety of sources—none are new. The umbrella concept of universal design brings these existing design strategies together in new ways, and that is the uniqueness and strength of the universal design approach.

The examples and applications associated with the strategies have been chosen to span as many engineering and design disciplines as possible. The universal design strategies derive from a simple premise: design entities to work with, support, and enhance human functioning. Design to human strengths and not human weaknesses. Create products and processes wherein people are allowed to do what they do best, and where the products and processes are required to do what they do best.

Chapter 5 provides an overview of the universal design principles. The design principles fall into three broad categories: (1) those dealing with human functions, (2) those dealing with processes, and (3) value-based design, which in practice places constraints on the human function and process principles. This hierarchical categorization of universal design principles is new to this book and affords a pragmatically useful way to organize the principles.

There are three chapters (Chapters 6 through 8) dealing with the human function principles of ergonomics, perception, and cognition. These three principles are those most often associated with universal design. The ergonomic considerations follow from traditional safety and risk reduction strategies as well as anthropometric factors. The perceptual issues start from psychomotor laws such as Weber's law, Fechner's law, and Steven's power law, which describe fundamental human sensory information processing and move to design strategies based on these laws. Dealing with cognitive issues is traditionally a very difficult problem and, consequently, a great deal of material is devoted to this very important topic. Chapter 8, which deals with cognitive issues, contains a wealth of cognitive design strategies.

There are four chapters (Chapters 9 through 12) dealing with process issues: flexibility, error-management, efficiency, and process variability and variation. Unique to this book is the discussion of the hierarchical linkages wherein the process-level design principles and strategies place constraints on the lower-level human functioning principles and strategies.

Flexibility and error-management are typically mentioned as universal design principles, but not from a process-oriented perspective. This book also expands on the traditional concept of flexibility to include the use of assistive technology and adaptive design strategies to achieve flexibility. These topics are important because they are explicitly mentioned in many of the laws on accessible design. This book uses the term "error-management" as distinct from error-proofing because not all design applications call for error-proofing, and hence error-management is a more accurate term.

The inclusion of efficiency and the reduction in process variability and variation as universal design process principles is unique to this book. Efficiency is characterized from the *kaizen* (Japanese continuous improvement methodologies) point-of-view of reducing or eliminating non-value-added activity. Non-value-added activities are defined and process-based universal design strategies are presented. Deming's ideas of common cause and special cause process variability form the theoretical basis for design strategies that seek to reduce the common cause variability of processes.

Chapter 13 covers value-based design, in particular, *equitable design* and its role as an integrating principle. Equitability is the last universal design principle considered in this discussion; its representation as the synergistic integration of the other universal design principles argues for its prominence in design requirements. This is why it is often the first cited universal design principle in other discussions.

Universal design traditionally focuses on creating nonstigmatizing, equitable designs. Nonstigmatizing and equitable imply being age and context appropriate, being aesthetically pleasing, being affordable, and having a broad market appeal. These design features and characteristics are true for any product or service, but are especially true for designers seeking to follow universal design principles.

What has not been recognized is the role of equitability as an integrating principle. A role made evident by the hierarchical structure among the design principles, equitability imposes constraints on the other design principles in that they must be applied so that the designed entities are accepted by a broad spectrum of users. The designs must be age and context appropriate as well as aesthetically pleasing. In a most fundamental way, equitability forces the integration of the other universal design principles.

Accessible design does not concern itself with being equitable. Accessible design is concerned with removing or preventing environmental barriers as prescribed by law, guidelines, and standards. This does not mean that an accessible design requirement cannot be equitable.

Chapter 14 takes a close look at universal design in the workplace, not so much specific strategies and methods in that these are presented in the previous chapters, but rather what makes the workplace unique from a broader social context. Examples are provided from large multinational corporations, small businesses venturing into the global marketplace, and a traditional non-profit service organization that is changing to meet the demands of e-commerce and global markets.

The final chapter of Section III, Chapter 15, presents the World Wide Web design community as a model and example of the mainstreaming of accessible and universal design into a globally organized design community.

Section IV consists of two chapters. This section provides a transition from the design of products and processes so as to enhance human functioning to viewing how the environment can contribute to defining a person's value. This transition is part of an international process of evolution and redefining of universal design. Chapter 16 explores the evolution of universal design. The chapter starts by examining the ethics of disability from both a personal and societal perspective. It then goes on to present three different definitions of universal design that exemplify the evolutionary trends and the ethical implications of these definitions. Chapter 17 returns to the themes of Chapter 1—that is, the rich and complex social and political context within which the practice of design takes place. It continues to explore the evolution of accessible and universal design from these societal and political perspectives. These perspectives enable us to step back and take a more reflective approach to accessible and universal design, and provide conclusions based on these reflections.

Author

Dr. Robert Erlandson is a professor of electrical and computer engineering and bioengineering at Wayne State University, Detroit, Michigan. Dr. Erlandson has a B.S. in electrical engineering from Wayne State University and a Ph.D. in biomedical engineering from Case Western Reserve University, Cleveland, Ohio. While Dr. Erlandson has had over thirty years of educational and R&D experience, his professional background is not that of your typical academic. He has worked at Bell Telephone Laboratories in the area of advanced switching systems, he was a partner and president of a small consulting firm, and served as vice president for research and technology development, for the Metropolitan Center for High Technology, a technology transfer and economic development organization operated jointly by Wayne State University and the State of Michigan. Currently he is director of the Enabling Technologies Laboratory in the College of Engineering.

Established in 1992, the mission of the Enabling Technologies Laboratory (ETL) is to design and develop technologies that enhance human capabilities and potentials, with a special emphasis on people with disabilities. The ETL is unique in its operations in that it combines community service, education and research. Through the ETL Dr. Erlandson has been able to channel his design, business, research, and educational experience and expertise to create a truly unique program.

Dr. Erlandson has authored over 100 technical papers and articles and has received numerous awards for his outstanding teaching and educational design activities. Most recently the American Society for Engineering Educators' awarded Dr. Erlandson the 2006 ASEE Fred Merryfield Design Award. This is a national award for exemplary and exceptional design programs. The Enabling Technology Laboratory program was awarded the Michigan Campus Compact 2000 Faculty/Staff Community Service-Learning Award. Michigan colleges and universities give this award to exemplary programs that integrate education, research and community service. Dr. Erlandson has received numerous awards for teaching excellence including Wayne State University's President's Award for teaching excellence.

Dr. Erlandson's professional background ranging from Bell Laboratories, to small business, to technology transfer and economic development, design, research, community service and education provides him with a unique mix of experience from which to draw when considering universal and accessible design for products, services, and processes.

section one

Introduction and definitions

Section goals:

- Present design as a societal process influencing and being influenced by societal elements.
- Provide a definition of universal and accessible design within the context of the general design process.
- Provide a rationale for the mainstreaming of accessible and universal design.
- Demonstrate that a gap exists in the design curricula of U.S. higher education with respect to accessible and universal design, current legal mandates, and global market conditions.

Scope of the discussion

The section goals are ambitious but are important in providing the broader societal context within which the notions of universal and accessible design are evolving. Design in its most general sense is a process practiced by virtually every profession. Two of the most commonly thought-of professions are engineering and architecture. However, universal design for learning is an example of the spread of universal design beyond its traditional disciplines. Teachers are designers, designing activities and curriculum material. Universal design for learning has emerged as the practice that covers the application of universal design principles in educational settings.

As universal design and its close first cousin, accessible design, evolve and their influence spreads, society changes its collective conceptions about human functioning. This process is driven by a collection of dynamic feedback loops among societal elements and design activities. A simple example illustrates the feedback processes.

A person who is blind, working as a telemarketer required to take orders via phone or computer and fill out electronic forms, can function as effectively as a sighted person with the appropriate technology—for example, a screen reader (providing text-to-voice output), a Braille printer. Without such technology, the blind worker would be disabled and unable to perform the job.

As a society, we either directly experience the beneficial effects of such technology, witness its impact first-hand, or hear stories about people who work with such technology, either from friends and colleagues or through the media. The cumulative effect of such experiences is a redefining of what it means to be disabled. Society's definitions of normalcy change to reflect the availability and utilization of such enabling and assistive technologies. And as has happened throughout the world, the availability of such technology and its witnessed impact have created laws mandating increased accessibility. While this is a very simple example, it illustrates the dynamic linkages and evolution among societal elements and design.

Chapter 1 systematically presents and discusses the aforementioned topics. The goal in crafting this chapter is to provide sufficient detail and examples so as to present a rational and reasonable argument for the mainstreaming of accessible and universal design into the various U.S. design communities. Chapter 1 reveals that universal and accessible design are driven by different yet complementary pressures, and that it is critical for American businesses to understand the different drivers so that they can appropriately and effectively respond to legally mandated accessibility versus market-driven forces to offer more accessible products and services.

Chapter 2 begins with a very general definition of the design process and then goes on to define universal design, accessible design, and adaptable design within that general design context. Following the definitions, a variety of examples are presented illustrating each definition and, more importantly, how the various design approaches can overlap one another.

chapter one

The time has come

Chapter goals:

- Present the global market pressures for the mainstreaming of universal and accessible design into the broader design community.
- Introduce the legal mandates for the mainstreaming of accessible design into the broader design community.
- Present the knowledge gaps that currently exist in U.S. higher education design programs.
- Present the current state of affairs with respect the mainstreaming of universal and accessible design into U.S. higher education design programs and hence the U.S. design community.

Linkages

Laws, global markets and competition, an aging population, technological advances, and evolving conceptualizations of disability and societal values are combining to promote the mainstreaming of universal and accessible design. Figure 1.1 diagrammatically illustrates the interactions of these forces. The process of design takes place within and this milieu. Design does not take place in isolation—it is a social process. It draws ideas, inspirations, and constraints from the social environment and, in turn, the products of design influence this environment. Design is placed at the center of Figure 1.1 because design is the focus of this book.

A variety of federal laws currently mandate the accessibility of buildings, transportation, telecommunication products and services, and electronic and information technology. These laws are major drivers behind the mainstreaming of accessible and universal design. The laws, derived from societal attitudes that grew out of the civil rights movements of the disability community, are based on the premise that the environment can "level the playing field" for people with disabilities.

Shortly after taking office, President George W. Bush created the *New Freedom Initiative* [1]. When announcing the New Freedom Initiative,

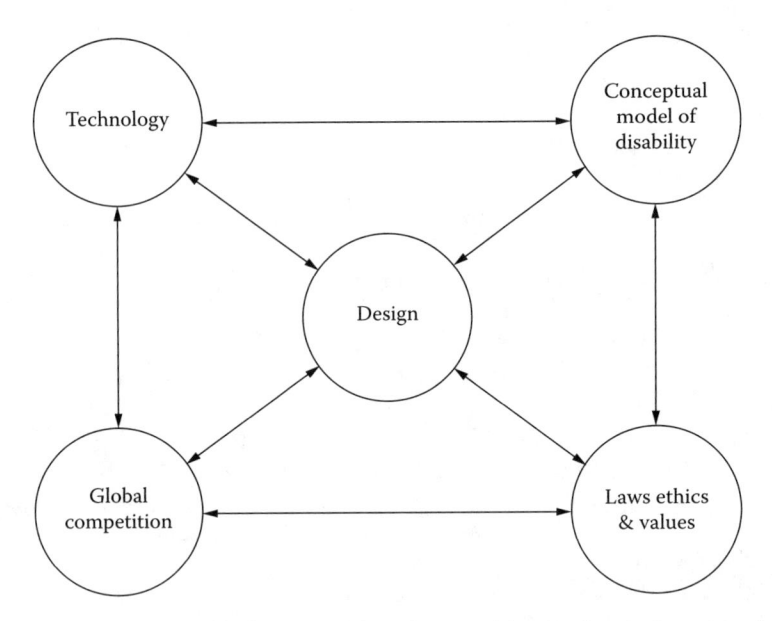

Figure 1.1 Technology, global competition, laws, ethics, and our conceptual model of disability all interact with one another and all influence the design process.

President Bush stated that "My Administration is committed to tearing down the barriers to equality that face many of the 54 million Americans with disabilities" [1]. This initiative is based on the premise that "[a]ssistive and universally designed technology offers people with disabilities better access than ever before to education, the workplace, and community life" [2].

Legal mandates, while initially the only major drivers for the mainstreaming of accessible and universal design, are being joined by an equally powerful set of forces, global competition, and global markets. Global competition and markets are now also driving the mainstreaming of universal and accessible design, perhaps more forcefully than legislation. Toyota's Universal Design Showcase in Tokyo, Japan, exemplifies the global trend toward universal design. This facility and associated website promotes and explains Toyota's corporate commitment to universal design in all of its operations, from automobiles to homes [3, 4].

A London-based company, OXO International, Ltd., exemplifies the success of mainstreaming the universal design into a company's design culture. OXO is a multinational company, with a global market for its products. OXO's product lines include cooking tools and utensils; serving and entertaining tools and equipment; cleaning, organizing, and gardening tools; utensils; and equipment. OXO was founded in 1990 on the philosophy of universal design, "which is the concept of designing products that are easy to use for the largest possible spectrum of users" [5]. OXO's *Who We Are* website states that "living by Universal Design principles gives us an opportunity to see things from a different perspective. The goal of making products

Chapter one: The time has come

Figure 1.2 The homepage website for Toyota's Universal Design Showcase. This facility showcases the benefits of universal design for everyone. (From TOYOTA Universal Design Showcase, http://www.megaweb.gr.jp/Uds/English/guide.html, 2004.)

more usable forces us to first identify problems and inefficiencies of existing products (including our own), not only in terms of comfort, but performance as well" [5].

OXO's marketing success also demonstrates that universally design products can be both functional and artful. Each year, OXO introduces more than 50 products [5]. OXO's designs have won about 100 design awards since 1991. These awards range from Kitchenware Winner 1991 (National Housewares Manufacturers Association, England) to the 1992 Industrial Design Excellence Award, Gold (Industrial Designers Society of America), to four Design Plus Awards in 2003 (Messe, Frankfurt, Germany). In addition to such industry recognition, OXO products were selected for inclusion in the permanent Design Collection of The Museum of Modern Art, New York, in 1994. See [6] for a complete list of OXO awards. These awards and recognition clearly demonstrate that accessible, universal design approaches not only address functionality, but can also be artistic and aesthetically pleasing, reaching a mass consumer market.

Definitions: Universal design and accessible design

For now, universal design can be defined as the design of facilities, products, and services that can be used by people of all abilities, to the greatest extent possible, without adaptations. This definition is a slight modification of the definition provided by the Center for Accessible Housing [7]. Accessible design can be defined as the design of facilities, products, and services that satisfy specific legal mandates, guidelines, or code requirements with the intent of providing accessibility to the entities for individuals with disabilities. This definition of accessible design focuses on the legal implications of the term. The definition is an expansion of the 1991 Center for Accessible Housing's definition [8].

Whereas accessible design is legally driven in that specific legal requirements must be met to ensure the accessibility of facilities, products, and services for individuals with disabilities, universal design is not legally driven or mandated. Quoting again from the OXO website: "The concept of Universal Design makes room for all users by taking as many needs as possible into consideration in the design process. It is important to note that Universal Design does not mean designing products fully usable by everybody, since there is no product that can truly fulfill the needs of all users. But when all users' needs are taken into consideration in the initial design process, the result is a product that can be used by the broadest spectrum of users" [9].

Laws

In the United States, laws have traditionally been and continue to be the dominant force for accessible design. The laws mandating accessible design reflect a broad shift in societal values and the conceptualization of disability. As people experience and witness the effect of technology and accessible/universal design to improve lives, the notion of disability is changing. Positive feedback of this sort ensures that change begets change. Expectations for one's own life and the lives of fellow citizens rise and this places pressure on the design process to further enhance products and services. Universal design is a logical conclusion to this positive feedback process.

Current federal laws are mandating that the built environment, buildings and facilities, transit vehicles, telecommunication products and services, and electronic information and technology be accessible for people with disabilities. These laws also created the Access Board, which is responsible for creating and publishing a collection of guidelines covering accessible design requirements for the built environment, transit vehicles, telecommunications equipment and services, and for electronic and information technology. All designers must be aware of the Access Board and the guidelines it publishes because these guidelines place constraints on the emerging design.

Table 1.1 provides a summary of the major laws and the affected disciplines. These laws, rules, and regulations impact all engineering design disciplines and cover the design spectrum from the assembly and manufacturing of products to consumer use of products and services. It is imperative that engineers, business owners, information/telecommunication system

Table 1.1 Major Laws and the Disciplines They Impact

Discipline/laws	Electrical & computer engineering	Mechanical engineering	Civil engineering	Industrial engineering	Chemical engineering	Biomedical engineering	Computer science	Engineering technology	Architecture	Business
Architectural Barriers Act			X						X	X
Rehabilitation Act[a]	X	X	X	X	X	X	X	X	X	X
Americans with Disabilities Act	X	X	X	X	X	X	X	X	X	X
Telecommunications Act, Section 255	X			X			X	X	X		X
Rehabilitation Act Amendments, Section 508	X	X		X				X	X		X

[a] The Rehabilitation Act created the Access Board—and hence impacts all areas.

designers, Web authors, and others be cognizant of the legal and social concerns that surround the issues of accessible design—issues that will have significant impact on the products, facilities, and services that such professionals provide.

Market and ethical imperatives for accessible and universal design

While laws drive accessible design, e-commerce and the global competition for products, services, and jobs are the dominant forces for universal design. Toyota and OXO exemplify the global pressures for mainstreaming universal design. In the United States, there are other market pressures that are also at work. A significant force arises from America's aging population, in particular the baby boomers. The baby boomers have redefined markets for products and services since their births and will continue to do so until they die. The baby boomers are beginning to experience the effects of aging, including increasing physical challenges, decreasing vision and hearing acuity, health conditions that affect stamina, and increasing cognitive challenges. In terms of their demand for more accessible products, services, and jobs, markets are beginning to feel and respond to the impact of this group.

An example of the significance of the baby-boomer market is Ford Motor Company's use of the "Third Age Suit" [10, 11]. This is a suit that inhibits movement, speed, and range of motion. It is used to allow young designers to experience the effects of aging and thereby sensitize them to the need for using accessible and/or universal design principles when designing cars and trucks. Another example is the agreement between the AARP and Home Depot for the employment of older individuals [12].

As already noted, global markets represent another force pushing for more universally designed products, services, and workplaces. Global markets driven by e-commerce and a technology base that supports global communication and transportation of products, services, and ideas make small businesses in Michigan direct competitors with similar small businesses in Italy, Brazil, or Japan. With such global competition and corresponding market base, businesses must be sensitive to and as responsive as possible to a variety of market niches.

Robert Reich, former Secretary of Labor under President Clinton, speaks to the "new economy" that the United States and the world now finds itself in [13], a global economy driven by e-commerce. Reich states two characteristics of the new economy. First, choices are widening for the consumer. Consumers can easily switch brands or supplier in search of a better deal. Second, the breadth of choices and relative ease of switching place suppliers and sellers in a very vulnerable position with respect to competition. This vulnerability drives innovation and the desire to create even more attractive products and services so as to retain and expand market share. However, as Reich points out, winning market share in the new economy is temporary—the race is never over [13].

Universal design is a natural design approach in such a volatile market environment. It is a cost-effective approach. Design, build, and sell products and services that as many people with as broad a spectrum of abilities can use without accommodation. For OXO, "the principles of Universal Design mean a salad spinner that can be used with one hand; liquid measuring cups that can be read from above without bending over; a toilet brush that bends to reach out-of-the-way places; a backlit oven thermometer that can be read easily through the window of an oven door; kettles with whistle lids that open automatically when tipped to pour; and tools with pressure-absorbing, non-slip handles that make them more efficient" [9].

Another example of the use of universal design due to global marketing is the use of a standardized icon collection for video and audio recorders and players, the stop, pause, record, forward, and reverse icons. Standardized icons in airports and other public places for toilets, exits, and food are yet further examples of universal design.

Accessible and universal design principles are natural strategies for the design of work and service processes to support a global distribution of work and the associated incredibly diverse workforce. Manufacturing operations must be responsive to the changing global market demands. These pressures demand faster prototyping and product development cycles, as well as agile assembly and manufacturing practices that support a mix of high- and low-volume orders [14, 15]. Design, manufacturing, assembly, and customer and service support functions can be distributed globally with a staggeringly diverse workforce. Transportable workplace and job designs that enable workers with different languages, different anthropometric features, and different skill levels to step into the process and with little or no training competently perform the required tasks are extremely desirable. Universal design in the workplace affords a design strategy aimed at creating transportable workplace designs that can effectively and efficiently serve the very diverse workforce found in a global economy.

As an example of universal design in the workplace, consider the ETL talking scale system [16]. A customized control unit with voice output and built-in recorder is interfaced to a digital scale. In addition to the scale's digital display (numeric) the controller has a three-light indicator (high, OK, low), and voice prompting. The voice prompting can be easily recorded and hence the prompts customized with respect to language and complexity. The controller must be configured for the weighing or counting task. If the voice prompt feature is turned on, the controller will inform the worker if the weight (or count) is "too high," or "too low," or "OK."

The talking scale system and controller were designed using universal design principles [16, 17]. Workers with cognitive impairments as well as non-impaired workers have used the device. For all workers, the controller/scale system yielded improved quality by reducing weighing and counting errors. The system's features worked differently for different workers [17]. For the cognitively impaired, it provided the support and feedback necessary to actually be able to do the job at the required performance levels. For the non-disabled worker, it provided error-proofing and improved quality control to fight boredom and fatigue.

The substantial disability market is another force driving the use of universal design. Accessible and universal design techniques provide a design strategy that inherently expands the market potential of a product or service by rendering it more accessible to a greater number of people. A market study of the U.S. population performed by the Electronic Industry Foundation states that "The National Center for Health Statistics (NCHS) and the Census agree that roughly 15 percent of the total population have some type of functional limitation. More than one-third of the total cannot perform major life activities like working, going to school, playing or caring for themselves" [18]. More significantly, research shows that "many people with disabilities are avid purchasers and users of products that help them communicate more effectively" [19]. It is an emerging realization that individuals with disabilities represent a potentially large market for a large variety of products and services [18–20].

The application of accessible/universal design principles provides a concrete and demonstrable ethical statement by business and industry. An example of industry's commitment is the June 3, 1998, SBC (Southwestern Bell Communication) policy on universal design (accessible design): "SBC's commitment to universal design principles is a tangible demonstration of the value SBC places on the worth and dignity of all individuals, including people with disabilities. SBC is committed to universal design" [21]. This statement is only one example of the ethical and compassionate efforts being shown in the business community for individuals with disabilities.

Hence, in addition to the pressures of globalization, there is a growing awareness that individuals with disabilities and the aging population represent significant market potentials for more accessible products and services. More and more businesses are recognizing the potential dollar value of what were once considered "niche markets." While the changing perceptions may have started as a rationalization and a coming-to-terms with the legal mandates, marketing to this particular consumer group is now recognized as smart, ethical business. This is a powerful combination.

Accessible and universal design: Mainstreaming issues

The legal mandates for accessibility, the market pressures for more universally designed products and services, and ethical concerns for ensuring the accessibility of products and services for people with disabilities, all speak to the mainstreaming of accessible and universal design into engineering and design. A study by the Enabling Technologies Laboratory (ETL) at Wayne State University in Detroit, Michigan, identified a lack of understanding of the legal and market issues as the biggest problem for greater mainstreaming [22].

Recent trends indicate that more people are becoming aware of the need for more universally designed products, services, and workplaces. Not surprisingly, the ETL study found that engineers working in areas most directly impacted by current laws and regulations that mandate accessibility were aware of the need for accessible and universal design, for example, those working on buildings and transportation systems, telecommunication systems, and electronic and information technology. Many of these practicing engineers have been involved with their company's market studies and have come to realize that there is a market potential for more accessible products.

The ETL study also confirmed that those individuals who are not working in areas directly affected by the laws mandating accessibility are less likely to know about the laws and are more inclined to think of universal design and accessible design as something nice to do for people with disabilities if development time and costs allow. These designers do not typically understand the broader global marketing implications for universally designed products and services. This situation will change. This book is but one example of the increasing availability of information about universal and accessible design.

The amount of publicly available information on accessibility laws, and accessible and universal design is rapidly growing. The federal government and those companies involved in complying with the laws and regulations have developed a significant amount of information on accessible design and its implementation, which is available online. The availability of such information is a driving force for mainstreaming, in that the increased availability of such material increases the likelihood that more people will discover the resources and utilize them. Table 1.2 provides some examples.

The European Union is aggressively pursuing the integration of universal design material in higher education as well as vocational and skilled crafts training programs. The Council of Europe, Committee of Ministers,

Table 1.2 Examples of Web-Based Material on Accessible Design

Example	URL
The U.S. Access Board	http://www.access-board.gov
Center for Applied Special Technology (CAST)	http://www.cast.org
TRACE Center	http://trace.wisc.edu
The Center for Universal Design	http://www.design.ncsu.edu:8120/cud
The Enabling Technologies Laboratory	http://www.ece.eng.wayne.edu/etl
Accessible Design Resource Center (ADRC)	http://www.etl-lab.eng.wayne.edu/adrc/Default.htm
IBM Accessibility Center	http://www-3.ibm.com/able
Microsoft Accessibility	http://www.microsoft.com/enable
W3C Web Accessibility Initiative (WAI)	http://www.w3.org/wai
Accessibility Forum	http://www.accessibilityforum.org/

stated in a recent resolution the need to integrate people with disabilities into the community using universal design or "design for all" principles [23]. A key component of their approach is to require education, training, and awareness for everyone dealing with the built environment [23]. With respect to higher education, this includes engineers, architects, designers, planners, and policy makers. The resolution then states that because a large number of vocations and skilled craftsmen are involved with designing and building in the physical environment, "… craftsmen, such as bricklayers, carpenters, plumbers, and electricians, the initial vocational training of all professions concerned should include universal design principles" [23].

In the United States, the Access Board is not only responsible for creating and publishing detailed accessibility guidelines related to the federal laws mandating accessibility, but is also responsible for technical assistance and training to help disseminate and ensure the design of accessible entities. As such, the Access Board has developed and made available technical assistance material and conducts a comprehensive training program for practicing engineers, designers, managers, and policy makers. The General Service Agency (GSA), the procurement arm of the federal government, and over 500 organizations (including corporations, professional organizations such as the IEEE, universities, and disability groups) have joined together and formed the Accessibility Forum. The Accessibility Forum was created to help both government and businesses more clearly define accessibility and evaluate whether electronic and information technology products and services are in compliance with the mandated accessibility requirements.

In summary, there are a multitude of organizations, both businesses and governmental, that are actively engaged in raising public awareness with respect to the needs of the disabled and the legal mandates requiring accessibility. As engineers and designers gain experience with universal and accessible design, they are discovering unexpected spin-off benefits and are beginning to rethink the marketing advantages of more universally designed products and services.

References

1. White House, "Fulfilling America's Promise to Americans with Disabilities," http://www.whitehouse.gov/infocus/newfreedom/, 2001.
2. White House, "The President's New Freedom Initiative for People with Disabilities: The 2004 Progress Report," http://www.whitehouse.gov/infocus/newfreedom/toc-2004.html, 2004.
3. Toyota, "TOYOTA Universal Design Showcase," http://www.megaweb.gr.jp/Uds/English/guide.html, 2004.
4. Toyota, "Toyota's initiatives in universal design," http://www.toyota.co.jp/en/environmental_rep/04/download/pdf/p67.pdf, 2004.
5. OXO, "Who We Are," http://www.oxo.com/about_whoweare.php, 2004.
6. OXO, List of awards for OXO products, http://www.oxo.com/about_awards.php, 2004.
7. Center for Accessible Housing, "Accessible Environments: Toward Universal Design," North Carolina State University, Raleigh, NC, 1995.
8. Center for Accessible Housing, "Definitions: Accessible, Adaptable, and Universal Design (Fact Sheet)," North Carolina State University, Raleigh, NC, 1991.
9. OXO, "What We're About," http://www.oxo.com/about_what.php, 2004.
10. Ergonomics Society Press Release, "Ergonomics in Car Design Benefits Older Drivers," 1999.
11. Loughborough University, "The Third Age Simulation Suit," 2001.
12. AARP, "Home Depot and AARP Launch National Hiring Partnership," http://www.aarp.org/Articles/a2003-11-12-scsep-homedepot.html, 2004.
13. R. Reich, *The Future of Success*. New York: Alfred A. Knopf, 2000.
14. J. P. Womack and D. T. Jones, *Lean Thinking*. New York: Simon & Schuster, Inc., 1996.
15. J. P. Womack, D. T. Jones, and D. Roos, *The Machine that Changed the World*. New York: Harper-Collins Publishers, 1990.
16. R. F. Erlandson and D. Sant, "Poka-Yoke Process Controller Designed for Individuals with Cognitive Impairments," *Assistive Technol.*, 10, 102–112, 1998.
17. R. F. Erlandson, "Accessible Design and Employment of People with Disabilities," in *The Sourcebook of Rehabilitation and Mental Health Practice*, D. Moxley and J. Finch, Eds. New York: Plenum, 2003, pp. 235–252.
18. EIF, *Extend Their Reach: Marketing to Consumers with Disabilities*. Arlington, VA: Electronic Industries Foundation, 1997.
19. E. Francik, *Telephone Interfaces: Universal Design Filter:* Human Factors Engineering, Pacific Bell, 1997.
20. E. Francik, *Computer & Screen-based Interfaces:* Universal Design Filter, Human Factors Engineering, Pacific Bell, 1997.
21. SBC, *SBC: Universal Design Policy*, Southwestern Bell Communications, 1998.
22. R. F. Erlandson and B. C. Babbitt, "The Movement of Accessible Design Principles into Mainstream Engineering: Now and Then," in *Emerging and Accessible Telecommunication and Information Technologies*, J. Winters, C. J. Robinson, R. C. Simpson, and G. C. Vanderheiden, Eds. Arlington, VA: RESNA Press, 2002, pp. 1–18.

23. Council of Europe Committee of Ministers, "Resolution ResAP(2001)1 on the introduction of the principles of universal design into the curricula of all occupations working on the built environment. Adopted by the Committee of Ministers on 15 February 2001, at the 742nd meeting of the Ministers Deputies," http://www.fortec.tuwien.ac.at/bk/BK-DOK-UnivDes.htm, 2001.

chapter two

Design: Universal design/ accessible design/adaptable design

Chapter goals:

- Provide definitions of universal design, accessible design, and adaptable design.
- Specify their relationship to the broader notion of design in general.
- Provide examples of the different design approaches.

What is design?

The word "design" can be either a noun or a verb. When used as a noun, *design* refers to an object or entity; and when used as a verb, *design* refers to a process [1]. For this discussion, *design* will be used as a verb, that is, design is a process. The Consummate Design Center [2], a virtual resource center for all types of designers, puts forth the following definition of design:

> Design is the thought process comprising the creation of an entity [2].

Each element of this definition provides insight as to the nature of design. Design starts with "thought," that is, insight, intuition, and reason [2]. Each of these elements will be considered. The designer must have insight, an idea, or a thought as to the connections between a design concept and the needs or problems addressed by the proposed entity. More simply, the designer needs to see the connection between "problem and possibility" [2].

Both Apple and Microsoft provide accessibility features in their respective operating systems. Designers at both corporations saw the link between the accessibility problems of people with disabilities and possible solutions, such as enlarged font and sound prompting for people with poor vision. These features demonstrate the integral link between problem and possibility that is crucial to the thought phase of effective design.

Intuition deals with deeper levels of knowing and derives from first-hand experience. Designers from both Apple and Microsoft worked with representatives from the disabled community to create designs that truly addressed real needs within that community. To be effective, the designer either has intuition based on personal experience or must work with others to "gain" the required intuition. For example, Ford Motor Company's Third Age Suit simulates many effects of aging [3]. Young designers are required to wear the suit in order to experience, to a limited degree, movement and perceptual limitations related to aging. As a result, designers have more insight as to the needs of elderly drivers and passengers.

Reason refers to the analytical methods of analysis: modeling, simulations, mathematics, and other systematic approaches to stating and comparing alternatives and differing design hypotheses. Computer operating systems go through a strict regime of simulation and testing. Automobile dynamics are thoroughly evaluated and tested on sophisticated computer simulation systems. Analytical methods are most often associated with engineering design processes.

Together, insight, intuition, and reason form the first crucial phase, the "thought" phase, of the design process.

The next component is *process*, that is, a collection of activities or steps. From a design perspective, the focus is on *thought processes* such as speaking, writing, drawing, building and testing models, simulations, etc. [2]. The process can be iterative or linear, simple or complex—there is no fixed process. For example, designing and building a simple wooden box can be a linear process: sketch or draw the box, collect the required materials, cut the wood, and assemble the parts. Designing a car is much more complex, so it is an iterative process: develop a proposed design, simulate the concepts identifying problem areas, then go back and modify the design. Designers will run through this cycle many times before finalizing the design.

Design is a thought process consisting of a variety of elements. "The whole of design comprises all the individual parts of that thought process leading up to, involved with, and even following the creation of the entity being designed" [2]. Depending on the entity being designed, these individual parts can include needs identification, alternative designs, modeling and/or simulations, prototyping, construction, quality control, sales, delivery, after-service, and customer feedback [2]. The design of an automobile comprises all of the elements just mentioned, from needs identification, through sales, delivery, and customer feedback.

Design leads to the creation of an *entity*. "That is, it leads to the tangible realization of a mature completion of the 'image of possibility' that originally served to initiate the process" [2]. The entity is the product of the design process. Where an entity might be a physical object (a car, computer), a temporal event or sequence of steps as with a play or a customer service process (taking an order), a conceptual formulation such as information theory or the theory of fuzzy sets, a relational formulation such as the operating procedures for a computer [2] or an environment (classroom, building, park, home, etc.).

As Miller notes, *"Design is the thought process comprising the creation of an entity"* [2]. As part of the "thought process," the designer needs to explicitly consider whether or not to use universal, accessible, or adaptable design principles. Such consideration derives from the perceived needs and specifications of the product or service and the designer's knowledge of universal, accessible, and adaptable design principles. The insight between problem and possibility needs to include universal, accessible, and adaptable design alternatives.

The definition of design put forth above covers not just engineering design, but design in its broadest sense. Within this broad conceptual category called *design*, more narrowly defined subcategories exist. Three such subcategories—universal design, accessible design, and adaptable design—will be considered in more detail.

Universal design

> Universal design can be defined as the design of entities that can be used and experienced by people of all abilities, to the greatest extent possible, without adaptations.

The above definition is a slight modification of the definition provided by the Center for Accessible Housing [4]. The modified definition uses the term "entity" instead of "products and environments" so as to convey a broader sense of application. Buildings [5], workplaces [6–8], products [9, 10], services [9], and both educational activities and materials [11] can all be universally designed. Curb cuts are the prototypical example of universal design. People in wheelchairs use curb cuts, but so do bike riders, skate boarders, and people pushing baby strollers. If universal design is properly conceived and implemented, it is not noticeable because it simply works.

The definition of universal design includes the notions of usability and accessibility. Usability and accessibility are not the same. Usability is multifaceted [12]. Typically, usability is expressed by five attributes [12]. First, the system should be easy to learn with users being able to quickly get started. Second, the system should be efficient in that once learned, the user can be productive. Third, the system should be easy to remember so that the casual user can return to the system after a short absence and use it without relearning. Fourth, the system should have a low error rate; that is, users should make few errors during use and easily recover from any errors that are made. Also, catastrophic errors must not occur. Finally, the system should be pleasant to use so that users are subjectively satisfied when they use it; that is, they like it. Implied in this expression of usability is that the system is accessible; that is, the user can physically access the system.

Usability and accessibility are related but different attributes of an entity (product or service). Usability implies accessibility in that if the user cannot physically access the system, then the system is, by default, not usable. Yet accessibility does not imply usability. For example, a person may be able to physically access a computer program, but it may be too difficult to learn and thereby be unusable.

Accessible design

> Accessible design is the design of entities that satisfy specific legal mandates, guidelines, or code requirements with the intent of providing accessibility to the entities for individuals with disabilities.

The above definition of accessible design focuses on the legal implications of the term. The definition is an expansion of the 1991 Center for Accessible Housing's definition [13], which referred only to "code requirements." As the legal environment has evolved, it is necessary to reflect the legal changes in the definition.

Accessible design derives its legal meaning from laws such as the Americans with Disabilities Act (ADA) [14], Section 255 of the Telecommunication's Act of 1996 [15], and Section 508 amendments to the Workforce Investment Act of 1998 [16]. The Access Board has published accessibility guidelines that provide specific design guidelines as related to each of the three laws mentioned [17]. These laws also mention that products should be compatible with assistive technology devices used by people with disabilities or able to be modified so as to be rendered accessible.

A design feature may be technically in compliance with the law but still be problematic. A blind mathematician reported that he has had to get down on his knees in some public elevators in order to read the Braille writing on the elevator controls. Because the controls were placed so low he could only read the Braille upside down while standing upright or even bending over. While being an ergonomically bad design, this also leads to a potentially embarrassing and stigmatizing situation.

Note that systems may be designed so that they are not accessible, but with specific modifications they can be made accessible to individuals with specific disabilities. Such modifications are termed "accommodations" and characterize the process of adaptable design.

Adaptable design

> Adaptable design features are modifications made to standard design for the purpose of making the design usable for an individual, as needed [13].

The above definition of adaptable design focuses on modifications made to existing entities that make the entity accessible to people with disabilities. Perhaps the most common example of adaptable design is van conversion that provides wheelchair accessibility for drivers and/or passengers. The vans are a standard design that is not wheelchair accessible; but after modifications are rendered, the vans are wheelchair accessible. The van accommodation is not required or mandated by any law, code, or guidelines; hence, it is not considered accessible design.

Therefore, adaptable design differs from accessible design in that laws do not mandate it, and it focuses on modifying an existing standard design. Adaptable design is not universal design because universal design creates products and services that are accessible and usable without adaptations.

The relationship of general design, universal design, and accessible design

The Venn diagram in Figure 2.1 illustrates the relationships among general design, universal design, accessible design, and adaptable design. This diagram is based on a similar diagram from Story [18], but differs from Story's diagram [18] in that universal design is not a subcategory of accessible design. This difference derives from the legal requirements associated with the definition of accessible design.

The rectangle labeled "Entities that are accessible" is explicitly used to emphasize the fact that accessibility can be achieved by design strategies from all design categories. If accessibility is achieved because it is mandated, then it results from accessible design strategies. If accessibility is achieved

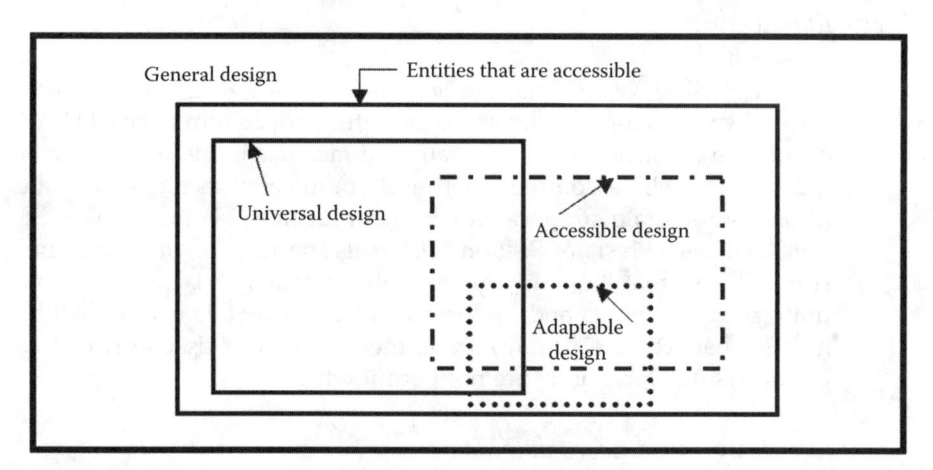

Figure 2.1 Venn diagram illustrating the relationship among general design, universal design, accessible design, and adaptable design.

as a result of an accommodation, it is an adaptable design strategy. Accessibility can also result from applying universal design principles.

Examples of entities from each design category

Universal design

1. *Adjustable seats, steering wheel, and floor pedals on cars and trucks.* The collection of adjustable features enable people with very different body sizes to use the car or truck. These adjustability features are not required or mandated by laws or regulations.
2. *Automatic door openers.* The automatic door openers at malls, shopping centers, and other public places are examples of universal design if laws or codes do not mandate them.

Accessible design and universal design

1. *Curb cuts.* Curb cuts were designed to allow wheelchair users to navigate and thereby have access to city streets and sidewalks, and the amenities they afford such as access to public places, buildings, shops, and restaurants. Curb cuts are mandated by ADA [14] and, as such, are an example of accessible design. However, curb cuts are used by people on roller blades, skateboards, bicycles, tricycles, people pulling shopping carts, and people pushing strollers, and therefore provide an example of universal design.

Accessible design

1. *The ADA Accessibility Guidelines for Buildings and Facilities* (ADAAG; http://www.access-board.gov/adaag/html/adaag.htm). The ADAAG covers the construction and alteration of facilities in the private sector (places of public accommodation and commercial facilities) and the public sector (state and local government facilities). For example, Section 4.9 deals with stairs. Section 4.9.2 deals specifically with treads and risers. It says that "[o]n any given flight of stairs, all steps shall have uniform riser heights and uniform tread widths. Stair treads shall be no less than 11 in. (280 mm) wide, measured from riser to riser (see Figure 18(a)). Open risers are not permitted."

Adaptable design and accessible design

1. *Volume controls for attaching to telephones.* A volume control that attaches to a telephone is an example of an adaptable design that is also accessible design. It is a modification made to a standard design

for the purpose of making it accessible to persons who are hard-of-hearing [18]. This design is accessible in that the Telecommunications Act of 1996, Section 255, requires telephones that do not contain volume controls to be compatible with technologies like the attachable volume control unit so as to be accessible for individuals with hearing difficulties [15].

Adaptable, accessible, and universal design

1. *Ramps to an existing public building when renovated*: When an existing public building undergoes major renovation, it may come under ADA accessibility requirements. One such adaptation is the addition of a wheelchair ramp. Ramps not only improve accessibility for people in wheelchairs, but also improve accessibility for people pushing strollers or pulling suitcases, or people with walkers. The ramps are like curb cuts in that they improve accessibility for all users and hence exemplify universal design. Ramps provided under the conditions described above are accessible designs because they are mandated. They are adaptable designs because they are added to a standard design.

Adaptable design

1. *Dampers for vibration reduction on tennis racquets*. One can purchase a variety of devices that fit around the strings of a tennis racquet near the frame. These devices change the vibration patterns of the racquet. If the undamped racquet produces a vibration pattern that causes undue stress on a player's elbow, it can result in "tennis elbow," which is a painful condition. The string damping device changes the racquet's vibration pattern so that it no longer places undue stress on the player's elbow and thereby avoids injury. The device is an adaptation in that it is a modification to a standard product that makes the racquet usable to an individual. For any given racquet, the adaptation only works for people with certain arm characteristics and not others; hence, it is not universal design. The device is neither mandated nor does it provide accessibility to individuals with disabilities; hence, it is not accessible design.
2. *Van conversions for wheelchair accessibility*. There is an active after-market business that centers on modifying vans to be wheelchair accessible for both the driver and passengers. This is adaptable design because a standard product is being modified so that it is accessible to individuals using a wheelchair. This is not an example of an entity in the intersection of adaptable and accessible design because no law, regulation, or code mandates van conversions.

Universal design in a global market

A 2002 report by the European Commission's Directorate—General Education and Culture, entitled *Teaching Universal Design: Global Examples of Projects and Models for Teaching in Universal Design at Schools of Design and Architecture*, summarizes universal design teaching activities in nine countries from Europe and Asia, and Australia and the United States [19]. It is clear that universal design takes on different meanings, depending on the economic development of a country. In the United States, Europe, and other economically developed countries, universal design is being driven by the desire to be more inclusive in particular with respect to people with disabilities. In underdeveloped countries, a major driver is the need is to use local materials and resources to develop products, services, and jobs that will sustain and help the local population to develop economically and educationally.

An example of the application of universal design in an underdeveloped country is the Freeplay radio [20], which was invented by Trevor Bayliss so that people living in southern Africa, who had little or no access to independent news and could not afford expensive radio equipment, could have access to independent news and public information. Bayliss invented an inexpensive wind-up radio that did not require batteries or a main power outlet. President Nelson Mandela invited him to set up a manufacturing plant in South Africa where the units are still being made. Figure 2.2 shows three different Freeplay radio models.

As evidenced by the three Freeplay models, the product not only serves the local target population, but also has evolved into a line of products that have become very popular with campers, hikers, and people who find themselves in isolated places without power or batteries.

(a) Eyemax

Figure 2.2 Three models of the Freeplay radio. The original Freeplay targeted people who could not afford a radio or did not have access to electrical power. Now, Freeplay has radios for a much broader market, including hikers and campers.

Figure 2.2 (a) Eyemax   Figure 2.2 (b) Devo   Figure 2.2 (c) Black.

Figure 2.2 (Continued)

(b) Devo

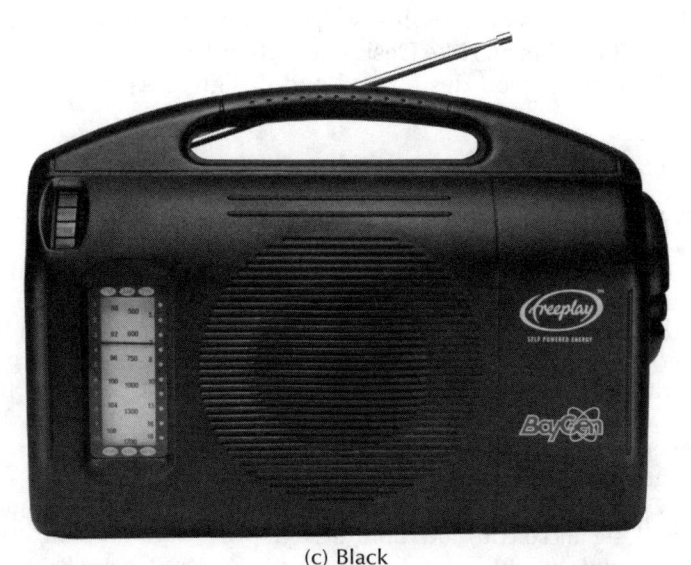

Figure 2.2 (Continued)

(c) Black

Conclusion

Design is the thought process comprising the creation of an entity. When conceptualizing the "problem and possibilities" from a universal design perspective, the possibilities include an entity that can be used and experienced by people of all abilities, to the greatest extent possible without adaptations. When approaching the same problem from an accessible design perspective, the possibilities will focus on how to satisfy legal mandates, guidelines, or codes to ensure accessibility. From an adaptable design perspective, the possibilities involve modifying or accommodating a standard design for the purpose of making it usable for an individual with a disability.

Highly competitive global markets and evolving accessibility legislation require corporations to maximize the market potential of their products and services while addressing the needs of individuals with disabilities. These dual forces require that designers understand and utilize different design perspectives. During the design thought process, designers should include a deliberate discussion and analysis of which design approach will be used to address the specified problem. Forgoing such deliberations could be very costly to a corporation in terms of lost market share, and possible litigation associated with a failure to comply with mandated accessibility requirements.

The European Concept for Accessibility Network makes the case for equating universal design with good design. In discussing diversity in the years 2000 and beyond, they urge Europeans to "no longer talk about the specific needs of certain categories of people, but talk about human functioning. We should look at every aspect of human functioning, without categorizing. ... Accessibility will lose its stigma and become a mainstream issue. We won't need terms like Design for All or Universal Design anymore. We will only refer to good design and bad design" [21].

References

1. Merriam-Webster, *Webster's Ninth New Collegiate Dictionary.* Springfield, MA: Merriam-Webster Inc., 1985.
2. W. R. Miller, "The Definition of Design," http://www.tcdc.com/dphils/dphil1.htm, 1996.
3. A. Lienert, "Mobility Challenges: Ford engineers age quickly in bid to woo older buyers. Outfit that simulates aging helps adapt designs to people with arthritis, bad backs and failing eyesight," in *The Detroit News*; http://detnews.com/2005/specialreport/0501/09/B06-54376.htm. Detroit, 2005.
4. Center for Accessible Housing, "Accessible Environments: Toward Universal Design," North Carolina State University, Raleigh, NC, 1995.
5. The Center for Universal Design, "Definition of Universal Design," http://www.design.ncsu.edu/cud/univ_design/princ_overview.htm, 2002.
6. R. F. Erlandson and D. Sant, "Poka-Yoke Process Controller Designed for Individuals with Cognitive Impairments," *Assistive Technol.*, 10, 102–112, 1998.
7. R. F. Erlandson, "Business and Legal Conditions Supporting the Employment of Individuals with Disabilities," in *The Sourcebook of Rehabilitation and Mental Health Practice*, D. Moxley and J. Finch, Eds. New York: Plenum, 2003, pp. 51–60.
8. R. F. Erlandson, "Accessible Design and Employment of People with Disabilities," in *The Sourcebook of Rehabilitation and Mental Health Practice*, D. Moxley and J. Finch, Eds. New York: Plenum, 2003, pp. 235–252.
9. TRACE, "Definition of Universal Design," http://trace.wisc.edu/docs/whats_ud/whats_ud.htm, 2002.
10. V. Fletcher and E. Steinfeld, "The World of Universal Design," IDEA Center/RERC on Universal Design at Buffalo, 2003.
11. R. F. Erlandson, "Universal Design for Learning: Curriculum, Technology, and Accessibility," in *ED-MEDIA 2002 World Conference on Educational Multimedia, Hypermedia & Telecommunication*. June 26–29, 2002, Denver, CO: Invited Paper.
12. J. Nielsen, *Usability Engineering*. San Francisco, CA: Morgan Kaufmann Publishers, 1993.

13. Center for Accessible Housing, "Definitions: Accessible, Adaptable, and Universal Design (Fact Sheet)," North Carolina State University, Raleigh, NC, 1991.
14. Access Board, "Revised ADA and ABA Accessibility Guidelines. Published in the *Federal Register* July 23, 2004 and amended August 5, 2005," http://www.access-board.gov/ada-aba/index.htm, 2005.
15. Access Board, *Telecommunications Act Accessibility Guidelines: Final Rule.* Washington, D.C.: Federal Register, 36 CFR Part 1193, 1998.
16. Access Board, *Electronic and Information Technology Accessibility Standards.* Washington, D.C.: Federal Register, 36 CFR Part 1194, 2000.
17. Access Board, "Access Board. This is the homepage URL for the Access Board. The Access Board is a key resource for accessible design issues," 2001.
18. M. F. Story, "Maximizing Usability: The Principles of Universal Design," *Assistive Technol.*, 10, 4-12, 1998.
19. B. Kennig and C. Ryhl, "Teaching Universal Design: Global Examples of Projects and Models for Teaching in Universal Design at Schools of Design and Architecture," AAOutils: ANLH, Brussels, 2002.
20. All That's Green, "The Freeplay Story," http://www.allthatsgreen.ie/page8.html, 2002.
21. European Concept for Accessibility Network, Web Home Page: European Concept for Accessibility Network (Text Sites: History, http://www.eca.lu/history.html; Definitions, http://www.eca.lu/def.htm), http://www.eca.lu/, 2001.

section two

Disability, laws, and accessible design

Section goals:

- Define disability within the context of three different conceptual models of disability: the medical model, the social model, and the systems model.
- Show how each conceptual model of disability has led to different design strategies and laws governing and mandating disability.
- Present the relationships between environmental factors and disability.
- Present the World Health Organization's new International Classification of Functioning, Disability, and Health as the basis for the book's conceptual model of disability.
- Present the major laws that have and continue to define accessible design in the United States.
- Introduce the Access Board and the collection of accessible design guidelines and training materials that, in essence, define accessible design in the United States.
- Introduce the Accessibility Forum and the resources and services it provides to businesses with respect to the compliance of accessible design requirements.

Scope of the discussion

Society's conceptualization of disability is perhaps the major motivating factor for society's laws governing accessibility and the treatment of people with disabilities. Society's conceptualization of disability influences the language used, and the balance between ethical and economic considerations found in its accessibility and disability laws. Section I began with an example of a person who is blind working as a telemarketer and served to demonstrate the connections between environmental factors (technology) and a person's ability to function in that environment. Figure 1.1 illustrated this dynamic linkage between technology and disability, but expands the picture by including global competition, and laws, ethics, and values. Chapter 3 explores in more detail the cumulative effect of such linkages in redefining what it means to be disabled. Society's definitions of normalcy change to reflect the availability and utilization of enabling and assistive technologies. The success of the enabling and assistive technologies to reduce disabling conditions, the new expectations of normalcy, and new laws mandating accessibility combine to influence the design process, challenging designers to design new technologies that outperform existing technologies.

Chapter 4 presents accessible design as a legally mandated activity. The chapter starts with a description of the Access Board. The Access Board is charged with developing and maintaining accessibility requirements as derived from federal laws, providing technical assistance and training on the guidelines and standards, and enforcing accessibility standards as prescribed by the law. To fully understand and appreciate accessible design, it is necessary to know about the Access Board and its functions.

Chapter 4 then provides a summary of the major U.S. laws dealing with accessibility and accessible design. Examples are provided showing the breadth and depth of these laws. These examples are also used to illustrate specific accessible design requirements and strategies.

Chapter 4 discusses and presents examples of the technical support and training provided by the Access Board. These are, in essence, very detailed and specific training and support materials for accessible design. Through its technical training and support functions, the Access Board is providing some of the best educational materials on accessible design. It behooves engineers and designers to take advantage of these excellent materials.

Chapter 4 concludes with a review of some of the international efforts on accessibility and accessible design. The United States is not alone in passing laws and developing accessibility standards. In a very pragmatic way, what the Access Board is presenting on accessible design is consistent with and required in much of the developed world.

chapter three

Disability and design

Chapter goals:

- Present a definition of disability.
- Present a conceptual model of disability that will be used throughout this book.
- Demonstrate how a conceptual model and definition can influence the design process.
- Discuss the implications for accessible and universal design.

Conceptual models

As society's conceptual model of disability has changed, corresponding changes in the laws regarding the disabled, their rights, and definitions of accessibility have occurred. Federal laws now mandate the accessibility of various products, facilities, and services. As these societal changes occur, the role of universal design becomes more prominent.

Conceptual models are formulations about system elements and the relationship between these elements that help people think about complex systems. For example, students can develop a conceptual model of how water can damage bridges during cold winters. Freezing rain or water formed from melting ice and snow may collect in cracks in concrete bridges. When the temperature drops below freezing and the collected water starts to freeze, the freezing water expands, increasing in volume. The freezing and expanding water exert a force on the concrete and expand the size of the crack. During the next thaw, more water can get into the crack; and when this water freezes, it expands the crack even further. This process eventually causes structural damage to the bridge.

This conceptual model contains the following specified system elements: water, concrete, cracks, temperature, and some interactions among these system elements. While this representation is incomplete based on partial knowledge, it allows students to think about the system elements and how they interact to cause structural damage. The results of this conceptual analysis are not quantitative, but rather qualitative. It has been said that the results of an analysis using conceptual models" are like an out-of-focus picture that partially reveals relationships" [1]. Engineers often use conceptual models for building hypotheses that can then be tested experimentally. As new knowledge and understanding are acquired, old conceptual models give way to newer versions.

Many conceptual models have been developed about disability, people, environments, social policy, and a host of related system elements. While such knowledge may be incomplete, designers can still use such conceptual models to think about interactions among the system elements and gain valuable insights.

Definition of disability

This book uses the definition of disability as found in the World Health Organization's new International Classification of Functioning and Disability (WHO-ICF) [2].

> Disability is characterized as the outcome or result of a complex relationship between an individual's health condition and personal factors, and of the external factors that represent the circumstances in which the individual lives [3, p. 17].

Disability is seen as an outcome of a relationship between the person and the environment. For example, if a person requires a wheelchair for mobility and entry into a building requires ascending steps, and no elevator or ramp exists, then the person in the wheelchair is disabled. If the building has a ramp and elevators, then the environment presents no barriers and the person is not disabled in that context. Likewise, if an office worker, who is blind, has technology at work such as a text-to-Braille printing system, text-to-voice software on her computer, a screen reader such as Jaws© or IBM's Home Page Reader©, and such technology allows her to perform her job as effectively and efficiently as a non-blind co-worker, then she is not disabled with respect to her work environment.

Figure 3.1 illustrates the major dimensions of the ICF classification and how they interact. For any person, his or her overall health condition and personal factors such as age, gender, education, religion, etc., will influence how he or she is able to function in any given context or environment. A person's body function and body structure are also essential elements. Body functions include memory; sensory functions (vision, hearing, taste, smell, touch); and cardiovascular, immunological, and respiratory functioning. Body structures include structures of the nervous system, eyes, ears, arms, hands, legs, toes, skin, digestive, and endocrine systems. It should be noted that while poor health (for example, cancer or diabetes) might imply a functional and/or structural problem, the converse is not generally true. For

Chapter three: Disability and design 31

```
┌─────────────────────────────────────────────┐
│ Environmental factors:                       │
│                                              │
│   ┌──────────────────────────────────┐      │
│   │ Person:                           │      │
│   │    ┌ ─ ─ ─ ─ ─ ─ ─ ─ ─ ┐         │      │
│   │      Health condition             │      │
│   │    │ (disorder or disease) │     │      │
│   │    └ ─ ─ ─ ─ ┬ ─ ─ ─ ─ ┘         │      │
│   │              ↕                    │      │
│   │    ┌──────────────────┐           │      │
│   │    │ Body function    │           │      │
│   │    │ and body structure│          │      │
│   │    └──────────────────┘           │      │
│   │              ↕                    │      │
│   └──────────────────────────────────┘      │
│       ┌─────────────────┐    ┌ ─ ─ ─ ─ ┐   │
│       │   Activities    │◄──►│ Personal │   │
│       │and participation│    │ factors  │   │
│       └─────────────────┘    └ ─ ─ ─ ─ ┘   │
│                                              │
└─────────────────────────────────────────────┘
```

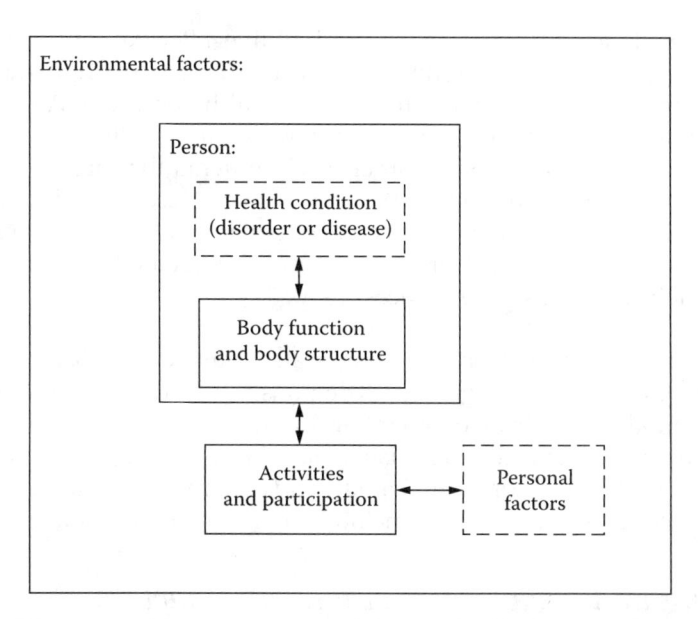

Figure 3.1 Diagrammatic representation of the WHO-ICF. A person functions (activities and participation) within an environment. The Person is characterized by his or her body function and body structure. The items designated by the dashed boxes are not explicitly covered within the WHO-ICF.

example, a person can be in good health and yet not be able to hear (function) because of an accident that damaged the eardrum (structure).

Activities and participation are essential elements of the ICF conceptual model. The two examples provided above describe two types of activities: getting into a building and office work. A person's level of participation depends on the environment. For the person in a wheelchair, no ramp or elevator (environmental factors) denies participation in activities that take place within the building. If the person's wheelchair (environmental factor) broke and was being repaired and he had no way to get another wheelchair, then his choice of activities and ability to participate would be severely limited.

For the office worker, the assistive technology in the work environment allows participation in a variety of work activities. If there were no assistive technology devices in the workplace, a person who is blind might be so disabled that participation in work activities is prohibited.

Focusing on the relationship between disability and environmental factors the ICF makes the following statement:

> Environmental factors interact with the components of Body Functions and Structures and Activities and Participation. ... Disability is characterized as the outcome or result of a complex relationship between an individual's health condition and personal factors, and of the external factors that represent the

circumstances in which the individual lives. Because of this relationship, different environments may have a very different impact on the same individual with a given health condition. An environment with barriers, or without facilitators, will restrict the individual's performance; other environments that are more facilitating may increase that performance. Society may hinder an individual's performance because either it creates barriers (e.g., inaccessible buildings) or does not provide facilitators (e.g., unavailability of assistive devices) [3, p. 17].

Environmental factors are very important from a design perspective; they are the things engineers design and, as demonstrated in the previous examples, environmental factors can be enabling or disabling. How engineers view human functioning and disability is critical to how they approach the design process. The WHO-ICF conceptual model and associated definition was chosen for this book because it explicitly includes environmental factors.

Enabling and disabling environmental factors

If an activity places excessively high functional demands on a person, then the person is disabled relative to those demands. For example, if a typical person is asked to drive a large truck that has no power steering, no power brakes, and a manual gear shift, that person probably could not do the job. Supplying the truck with power steering, power brakes, and an automatic gear shift allows most people to drive the truck. The power steering, power brakes, and automatic gear shift are environmental enablers because they enable a person's functional capability (ability to participate) for driving a truck (activity).

A replacement hip joint, eyeglasses, hearing aids, artificial hearts, prosthetic limbs, or orthotic devices such as knee or wrist braces, are also environmental enablers. These enablers work to restore a person's functional capabilities, that is, functional restoration. Functional restoration can also come about through exercise, training, and physical or occupational therapy.

From a design point-of-view, one does not want to design devices, products, or services that are disabling because the design prevents people from participating in an activity.

Design implications associated with the WHO-ICF conceptual model

The WHO-ICF model is an integration of two earlier conceptual models of disability: (1) the medical model and (2) the social model. Each of these models influenced laws and social policy, thereby influencing how engineers and designers approached designing products and services for people with disabilities.

Medical model

In the medical model, disability is viewed as a problem of the person caused by injury, disease, or some other health condition; the disability is treated by fixing the person, correcting the deficit. Physicians, physical and occupational therapists, or other rehabilitation service providers provide treatment. Management of disability is targeted at curing the person or at therapy that helps the person adjust to his or her disability. At the political level, the objective is to modify or reform health care policy so as to be more responsive to individual needs [1, 3, 4]. The 1980 International Classification of Impairments, Disabilities, and Handicaps (ICIDH), by the World Health Organization [5] exemplifies the medical model. Table 3.1 shows the major elements of the 1980 model.

Social model

The social model sees disability as a socially created problem with its central focus being the full integration of individuals into society. The social model has grown out of the disabilities rights movement. In 1976, the Union of the Physically Impaired against Segregation (UPIAS) published the *Fundamental Principles of Disability*. This is one of the earliest and clearest articulations of the fundamental principles of what has come to be known as the social model:

> ...disability is a situation, caused by social conditions, which requires for its elimination, (a) that no one aspect such as incomes, mobility or institutions is treated in isolation, (b) that disabled people should, with the advice and help of others, assume control over their own lives, and (c) that professionals, experts and others who seek to help must be committed to promoting such control by disabled people [6].

Table 3.1 The 1980 International Classification of Impairments, Disabilities, and Handicaps by the World Health Organization

1980 ICIDH
Disease: Something abnormal within the individual; etiology gives rise to change in structure and functioning of the body.
Impairment: Any loss or abnormality of psychological, physiological, or anatomical structure or function at the organ level.
Disability: Any restriction or lack (resulting from an impairment) of ability to perform an activity in the manner or range considered normal for a human being.
Handicap: A disadvantage resulting from an impairment or disability that limits or prevents fulfillment of a normal role, depending on age, sex, and sociocultural factors.

Source: From WHO, International Classification of Impairments, Disabilities, and Handicaps, World Health Organization, Geneva, Switzerland, 1980.

Systems model

The WHO-ICF model embodies elements of what can be called a systems model of disability. The systems model sees disability as a broadly based distributed phenomenon with its central focus that of maximizing an individual's functional capabilities within a spectrum of environmental conditions. The systems model integrates elements from both the medical and social models.

For the systems model, as with the medical model, there is a focus on the person; but drawing from the social model, the person is an active agent rather than a passive agent for whom and to whom things are done. In the systems model, people must assume responsibility for those elements over which they can exercise some control (for example, lifestyle, diet, exercise, and utilization of the resources available to them in their respective environmental conditions). These environmental resources include schools, clinics, businesses, transportation, social services, and insurance. Cultural and societal actions are concurrently required to ensure the availability and accessibility of the resources and opportunities necessary for an individual to maximize his or her functional capabilities within a given environment.

In the systems model, as with the social model, disability is not seen as an attribute just of the person, but rather brought about by a complex set of interactions with the environment. Environment is characterized by everything outside the person. This includes the physical environment and its components the natural and built environments, and also the cultural and social environments with all their associated components. From this perspective, the management of disability requires harmonious actions from both the person and the environment. Political actions therefore focus on ensuring the accessibility of resources and the removal of barriers, but also on promoting a variety of alternatives for personal actions and personal choices.

Accessible design, universal design, and conceptual models of disability

The medical model, social model, and systems model of disability suggest different roles and emphasis on design and the design process. The medical model views disability as a problem of the person and, as such, the notions of accessible and universal design are not relevant because these design approaches cannot restore the person's functional ability. The medical model's engineer would emphasize designing and developing prosthetics and orthotics that can directly restore a person's limited or lost functionality. Advances in bioengineering, materials, and nano-technologies are bringing about a resurgence in prosthetic and orthotic research, design, and development. However, these new engineering efforts in functional restoration will be part of a broader-based philosophy that emphasizes a spectrum of enabling activities.

Social models recognize the importance of the environment in defining disability. The various social models advocate using universal design to reduce or remove accessibility barriers. Laws that emerged under the influence of social models mandate the accessibility of facilities [7], products, and services [8, 9]. In many cases, the laws mandate the use of accessible design principles by providing specific design guidelines in the body of the law and/or require the development of accessibility guidelines for designers [8–10].

The systems model is similar to the social model in its advocacy of accessible design. However, the systems model places a stronger emphasis on using universal design. Universal design is seen as a way to increase accessibility for everyone. The systems model emphasizes the delivery of more accessible products from all segments of the economy, even those not driven by the legal requirements for accessibility. For example, the aging baby boomers are and will be demanding more accessible products and services. Also, the demands of global markets and an incredibly diverse consumer base require an increased use of universal design.

The evolution toward the systems model has elevated the importance of accessible and universal design as tools to reduce disability induced by the physical environment. Elements of the social environment, laws, public policy, public opinion, values, and market forces are combining to require or encourage utilizing accessible and universal design strategies. This is evidenced by the language used to describe the basic premise of the *New Freedom Initiative*: "Assistive and universally designed technology offers people with disabilities better access than ever before to education, the workplace, and community life" [11].

Conceptual model for this book

The World Health Organization's ICF will be used as the conceptual model of disability for this book [3]. Disability will be seen "as the outcome or result of a complex relationship between an individual's health condition and personal factors, and of the external factors that represent the circumstances in which the individual lives" [3, p. 17].

This conceptualization and definition have significant implications with respect to the design of entities because they force designers to constantly inquire as to the enabling or disabling pressures inherent in the emerging entity.

References

1. E. N. Brandt and A. M. Pope, "Enabling America: Assessing the Role of Rehabilitation Science and Engineering," Washington, D.C.: National Academy Press, 1997.
2. WHO, "Towards a Common Language for Functioning, Disability and Health: ICF," World Health Organization, Geneva, 2002.

3. WHO, "ICF Introduction," http://www.who.int/classification/icf/intros/ICF-Eng-Intro.pdf, 2001.
4. NIDRR, "National Institute on Disability and Rehabilitation Research (NIDRR) Long Range Plan for Fiscal Years 1999–2004," *Federal Register*, 63, 57910–57219, 1998.
5. WHO, "International Classification of Impairments, Disabilities, and Handicaps," World Health Organization, Geneva, Switzerland, 1980.
6. Union of the Physically Impaired against Segregation, "Fundamental Principles of Disability," London, 1976.
7. Access Board, "Americans with Disabilities Act Accessibility Requirements," http://www.access-board.gov/bfdg/adares.htm., 1999.
8. Access Board, *Telecommunications Act Accessibility Guidelines: Final Rule*. Washington, D.C.: Federal Register, 36 CFR Part 1193, 1998.
9. Access Board, *Electronic and Information Technology Accessibility Standards*. Washington, D.C.: Federal Register, 36 CFR Part 1194, 2000.
10. Access Board, "ADA Accessibility Guidelines for Buildings and Facilities (ADAAG)," http://www.access-board.gov/adaag/html/adaag.htm, 2002.
11. White House, "The President's New Freedom Initiative for People with Disabilities: The 2004 Progress Report," http://www.whitehouse.gov/infocus/newfreedom/toc-2004.html, 2004.

chapter four

Accessible design

Chapter goals:

- Inform the reader of the Access Board and its responsibilities.
- Present the major laws mandating the accessibility of buildings, facilities, transportation systems, telecommunication products and services, and electronic and information technology.
- Relate the laws and Access Board functioning to accessible design.
- Inform the reader of the training and educational resources distributed by the Access Board.
- Inform the reader of the Accessibility Forum and its resources and activities.

Access Board

The Access Board creates and publishes accessibility guidelines and standards mandated in various laws. These guidelines and legally mandated standards are, in practice, a statement of accessible design strategies and methods. Because of this, the Access Board is the focal point for information, training, and enforcement of accessibility-related issues. Knowledge of the Access Board, its duties and responsibilities, but more importantly its resources, is critical for engineers, designers, managers, and policy makers.

Section 502 of the Rehabilitation Act of 1973 created the Architectural and Transportation Barriers Compliance Board, which has come to be called, more simply, the Access Board or Board. At its creation, the Access Board was charged with ensuring federal agency compliance with the Architectural Barriers Act (ABA) of 1968 and proposing solutions to the environmental barriers and problems addressed in the ABA. Recent legislation such as the Americans with Disabilities Act (ADA), the Telecommunications Act of 1996, and Section 508 of the Rehabilitation Act have all required the Access Board to provide accessibility guidelines and training to help government and businesses comply with the mandated accessibility requirements.

The Access Board is an independent federal agency dealing with accessibility for people with disabilities. Key responsibilities of the Board include [1]:

- Developing and maintaining accessibility requirements for the built environment, transit vehicles, telecommunications equipment, and for electronic and information technology
- Providing technical assistance and training on these guidelines and standards
- Enforcing accessibility standards for federally funded facilities

The Access Board has a staff of about 30 people and a governing board of representatives from 12 federal departments. These are:

1. Department of Health and Human Services
2. Department of Transportation
3. Department of Housing and Urban Development
4. Department of Labor
5. Department of the Interior
6. Department of Defense
7. Department of Justice
8. General Services Administration
9. Department of Veterans Affairs
10. United States Postal Service
11. Department of Education
12. Department of Commerce

The governing board also has 13 public members appointed by the president of the United States, the majority of which are individuals with disabilities [2].

The Access Board is charged with training designers about accessibility laws and guidelines and providing technical assistance in applying these guidelines [3]. The Access Board conducts training sessions at sites across the country. The training sessions are typically customized to the needs of the audience. In a typical year, the Board conducts about 75 training sessions, reaching about 9000 people, in about half the states [4]. The Board posts its training schedule on its training website; see http://www.access-board.gov/research&training/Training.htm for the current schedule.

The first priority of the Access Board is to develop and maintain accessibility guidelines pursuant to federal laws mandating accessibility. Table 4.1 summarizes some of the currently available Access Board published guidelines. The major laws mandating accessibility and the creation of these guidelines are discussed next.

Chapter four: Accessible design

Table 4.1 Access Board Published Guidelines and Standards

Guideline or standard	Description of material covered
1. Uniform Federal Accessibility Standards (UFAS)	This document sets standards for facility accessibility by physically handicapped persons for federal and federally-funded facilities. These standards are to be applied during the design, construction, and alteration of buildings and facilities to the extent required by the Architectural Barriers Act of 1968, as amended.
2. ADA Accessibility Guidelines (ADAAG)	This document contains scoping and technical requirements for accessibility to buildings and facilities by individuals with disabilities under the Americans with Disabilities Act (ADA) of 1990. These scoping and technical requirements are to be applied during the design, construction, and alteration of buildings and facilities covered by Titles II and III of the ADA to the extent required by regulations issued by federal agencies, including the Department of Justice and the Department of Transportation, under the ADA.
3. State and Local Government Facilities: ADAAG Amendments	The Architectural and Transportation Barriers Compliance Board (Access Board) is issuing final guidelines to provide additional guidance to the Department of Justice and the Department of Transportation in establishing accessibility standards for new construction and alterations of state and local government facilities covered by Title II of the Americans with Disabilities Act (ADA) of 1990. The guidelines will ensure that newly constructed and altered state and local government facilities are readily accessible to and usable by individuals with disabilities in terms of architecture, design, and communication. The standards established by the Department of Justice and the Department of Transportation must be consistent with the guidelines.
4. Building Elements for Children: ADAAG Amendments	The Architectural and Transportation Barriers Compliance Board (Access Board) is issuing final guidelines to provide additional guidance to the Department of Justice and the Department of Transportation in establishing alternate specifications for building elements designed for use by children. These specifications are based on children's dimensions and anthropometrics and apply to building elements designed specifically for use by children ages 12 and younger. This rule ensures that newly constructed and altered facilities covered by Titles II and III of the Americans with Disabilities Act of 1990 are readily accessible to and usable by children with disabilities. The standards established by the Department of Justice and the Department of Transportation must be consistent with these guidelines.

Table 4.1 Access Board Published Guidelines and Standards (Continued)

Guideline or standard	Description of material covered
5. Play Areas: ADAAG Amendments	The Architectural and Transportation Barriers Compliance Board (Access Board) is issuing final accessibility guidelines to serve as the basis for standards to be adopted by the Department of Justice for new construction and alterations of play areas covered by the Americans with Disabilities Act (ADA). The guidelines include scoping and technical provisions for ground level and elevated play components, accessible routes, ramps and transfer systems, ground surfaces, and soft contained play structures. The guidelines will ensure that newly constructed and altered play areas meet the requirements of the ADA and are readily accessible to and usable by individuals with disabilities. The Department of Justice must adopt the guidelines as standards for them to be enforceable under the ADA.
6. Recreation Facilities: ADAAG Amendments	The Architectural and Transportation Barriers Compliance Board (Access Board) is issuing final accessibility guidelines to serve as the basis for standards to be adopted by the Department of Justice for new construction and alterations of recreation facilities covered by the Americans with Disabilities Act (ADA). The guidelines include scoping and technical provisions for amusement rides, boating facilities, fishing piers and platforms, golf courses, miniature golf, sports facilities, and swimming pools and spas. The guidelines will ensure that newly constructed and altered recreation facilities meet the requirements of the ADA and are readily accessible to and usable by individuals with disabilities.
7. ADA Accessibility Guidelines for Transportation Vehicles	This part provides minimum guidelines and requirements for accessibility standards to be issued by the Department of Transportation in 49 CFR part 37 for transportation vehicles required to be accessible by the Americans with Disabilities Act (ADA) of 1990, 42 U.S.C. 12101 et seq.
8. Over-the-Road Buses: ADA Accessibility Guideline Amendments	The Architectural and Transportation Barriers Compliance Board and the Department of Transportation amend the accessibility guidelines and standards under the Americans with Disabilities Act for over-the-road buses (OTRBs) to include scoping and technical provisions for lifts, ramps, wheelchair securement devices, and moveable aisle armrests. Revisions to the specifications for doors and lighting are also adopted. The specifications describe the design features that an OTRB must have to be readily accessible to and usable by persons who use wheelchairs or other mobility aids. The Department of Transportation has published a separate rule elsewhere in today's *Federal Register* which addresses when OTRB operators are required to comply with the specifications.

Chapter four: Accessible design

9. Telecommunications Act Accessibility Guidelines	The Architectural and Transportation Barriers Compliance Board (Access Board or Board) is issuing final guidelines for accessibility, usability, and compatibility of telecommunications equipment and customer premises equipment covered by Section 255 of the Telecommunications Act of 1996. The Act requires manufacturers of telecommunications equipment and customer premises equipment to ensure that the equipment is designed, developed, and fabricated to be accessible to and usable by individuals with disabilities, if readily achievable. When it is not readily achievable to make the equipment accessible, the Act requires manufacturers to ensure that the equipment is compatible with existing peripheral devices or specialized customer premises equipment commonly used by individuals with disabilities to achieve access, if readily achievable.
10. Final Standards for Electronic & Information Technology	The Architectural and Transportation Barriers Compliance Board (Access Board) is issuing final accessibility standards for electronic and information technology covered by Section 508 of the Rehabilitation Act Amendments of 1998. Section 508 requires the Access Board to publish standards setting forth a definition of electronic and information technology and the technical and functional performance criteria necessary for such technology to comply with Section 508. Section 508 requires that when federal agencies develop, procure, maintain, or use electronic and information technology, they shall ensure that the electronic and information technology allows federal employees with disabilities to have access to and use of information and data that is comparable to the access to and use of information and data by federal employees who are not individuals with disabilities, unless an undue burden would be imposed on the agency. Section 508 also requires that individuals with disabilities, who are members of the public seeking information or services from a federal agency, have access to and use of information and data that is comparable to that provided to the public who are not individuals with disabilities, unless an undue burden would be imposed on the agency.

Source: See http://www.access-board.gov/indexes/pubsindex.htm.

Laws and accessibility

The laws are presented in chronological order and as such will reveal the evolution, in terms of public thoughts and values, of society's conception of disability. Reference is made to the various conceptual models of disability discussed in Chapter 3 and the relationship to universal and accessible design also is presented.

Architectural Barriers Act (ABA) of 1968 as amended 42 U.S.C. § 4151 et seq.

The ABA requires that facilities built, altered, or leased with federal funds are accessible to individuals with disabilities; this is one of the first efforts by Congress to ensure access to the built environment. The ABA is a milestone in that it recognizes the disabling features of the environment and seeks to ensure that future federal facilities remove accessibility barriers.

The 1978 amendments to Section 502 of the Rehabilitation Act of 1973 (29 U.S.C. § 792) authorized the Board to establish minimum accessibility guidelines under the ABA and to ensure compliance with the requirements. The Board was directed to provide technical assistance and provide help on the removal of barriers including, for the first time, communication barriers in federally funded buildings and facilities. For the first time the Board was also directed to provide technical assistance to private entities to the extent practicable.

In compliance with this law, the Board has prepared and published the Uniform Federal Accessibility Standards (UFAS) (see item 1, Table 4.1). The UFAS sets standards for facility accessibility by physically handicapped persons for federal and federally funded facilities [5]. These standards should be applied during the design, construction, and alteration of buildings and facilities to the extent required by the ABA. These standards specify what accessible design means relative to the ABA.

1990 Americans with Disabilities Act (ADA)

The ADA deals with accessibility in very broad terms, prohibiting discrimination on the basis of disability in employment (Title I), state and local government services and transportation (Title II), public accommodations and commercial facilities (Title III), and telecommunications (Title IV) [6]. The ADA covers a wide range of disabilities, from physical conditions affecting mobility, stamina, sight, hearing, and speech, to conditions such as emotional illness and learning disorders.

Title I

Title I of the ADA covers a number of accessibility issues related to employment. The ADA prohibits employment discrimination against qualified individuals who have a disability. The ADA defines "a qualified individual with a disability" as one able to meet legitimate skill, experience, education, or other requirements of employment and who can perform the "essential functions" of the job with or without a "reasonable accommodation" [6]. The law goes on to define "a reasonable accommodation" as any modification or adjustment to a job or the work environment that will enable a qualified applicant or employee with a disability to perform the essential functions of the job [6].

A reasonable accommodation might be assistive technology and the associated assistive technology service. An *assistive technology device* is defined as "any item, piece of equipment, or product system that is used to increase, maintain, or improve functional capabilities of individuals with disabilities" [6]. An *assistive technology service* is defined as "any service that directly assists an individual with a disability in the selection, acquisition, or use of an assistive technology device" [6].

Title I views disability from a social model perspective in that the intent is to assist the person either with a reasonable accommodation or assistive technology. Both are environmental interventions aimed at helping the person with the disability. A systems model perspective would encourage using universal design strategies in the workplace; that is, use design principles that reduce both the physical and cognitive demands of the job, thereby making it safer for everybody, while improving quality and productivity for all workers. Such an approach fundamentally changes the "essential functions" of a job and thereby redefines a "reasonable accommodation" as well as the capabilities required by a qualified applicant.

Erlandson has shown that using such universal design strategies in the workplace does indeed address businesses' need for improved quality and productivity, while concurrently creating job opportunities for people with disabilities [7–9]. One very dramatic example deals with the assembly of a fuel filter clamp. The job consisted of placing adhesive back pads onto fuel filter clamps with very little margin for error [7]. Non-disabled workers at the company who manufactured the clamps could not produce assembled clamps at the required productivity level. These workers were frustrated, and management sought to outsource the assembly job.

A special education training workshop bid on and was awarded the fuel filter clamp assembly job. The assembly fixture, used by workers at the manufacturing company, was supplied to the workshop as an assembly aid. This fixture allowed for placing the metal clamp on the supporting structure in several orientations. Improper placement of the clamp increased the likelihood of improper placement of the rubber pad. Also, the metal clamps as supplied by the manufacturer exhibited considerable variation. Hence, even with the clamp placed in the proper orientation, movement of the clamp hindered accurate placement of the sticky-backed pad. A visual inspection was the only checking used to determine if the rubber pad was properly placed. If the pad was not properly placed, it had to be removed and reapplied. This was a time-consuming task that severely limited daily production rates. Staff and special education workers tried working with the supplied fixture and no one, non-impaired or cognitively impaired, could produce at the required level. The original fixture design made it difficult for everyone to perform this assembly without errors.

From a universal design point of view, the fixture promoted errors rather than hindered them. The fixture introduced considerable variation into the assembly process. While this inherent variation made it difficult for all workers to perform the assembly task, it made it virtually impossible for the workers having cognitive impairments. The high degree of variability and error promotion features made any kind of standardized work procedure impossible. The workers tried to avoid this task. They were not pleased with their own work. They tried very hard to perform the task but the job process itself was too much to overcome.

The fixture was redesigned using universal design principles to accommodate the variability of supplied blank clamps and yet still allow for accurate placement of the pad. Data comparing production rates and error rates for individuals with cognitive impairments between use of the old fixture and the new showed that the production rate increased by about 80% and the error rate dropped from above 50% to about 1%. These are dramatic results. More significantly, a large number of individuals who could not perform the assembly task with the old fixture were able to competently perform the task with the new fixture [7]. The special education training workshop retained the assembly contract as long as the fuel filter clamp was produced. By applying universal design, the special education workers exceeded specified quality and productivity requirements.

The redesigned clamp assembly fixture could not be considered a reasonable accommodation or assistive technology under Title I of the ADA because the fixture improved the job performance of all workers and was not targeted solely at the cognitively impaired. Not being considered a reasonable accommodation or assistive technology means that an employer cannot receive tax credits or other tax-related inducements for hiring individuals with disabilities if universal design strategies were utilized as the basis for creating jobs and hiring people with disabilities. The language of the ADA, Title I, does not support a systems model conceptualization of disability.

Titles II and III

Under Titles II and III, the ADA requires the Board to "issue guidelines to ensure that buildings, facilities, and vehicles covered by the law are accessible, in terms of architecture and design, transportation, and communication, to individuals with disabilities" [10]. These guidelines are called the Americans with Disabilities Act Accessibility Guidelines (ADAAG). The guidelines cover buildings, transportation (rail, air, buses, boats), building elements for children, play areas, and recreational facilities. The guidelines are very specific and detailed.

All the ADA-related guidelines (items 2 through 8 in Table 4.1) provide both breadth and depth with respect to the range of factors considered and

Chapter four: Accessible design

the detail with which the accessibility requirements are presented. Adherence to these respective guidelines constitutes accessible design. Titles II and III follow from a social model perspective. The intent is to remove existing environmental barriers and not introduce new barriers with new designs and construction.

The following examples are taken from *ADA Accessibility Guidelines for Buildings and Facilities (ADAAG)*. This document contains technical requirements for accessibility to buildings and facilities by individuals with disabilities under the ADA. These technical requirements are to be applied during the design, construction, and alteration of buildings and facilities covered by Titles II and III of the ADA [11].

Section 4.1.3, "Accessible Buildings: New Construction," deals with all aspects of accessibility including such topics as doors, ramps, entrances, parking spaces, water closets, toilets, alarms, signage, telephones, assembly areas, automated teller machines, dressing and fitting rooms, saunas and steam rooms, and benches.

Section 4.13 deals with doors. For example, 4.13.5 Clear Width: doorways shall have a minimum clear opening of 32 in. (815 mm) with the door open 90°, measured between the face of the door and the opposite stop. Additionally, 4.13.7 Two Doors in Series: the minimum space between two hinged or pivoted doors in series shall be 48 in. (1220 mm) plus the width of any door swinging into the space. Doors in series shall swing either in the same direction or away from the space between the doors. Section 4.13.11 specifies the "Door Opening Force" for a variety of door types.

Section 4.15 deals with "Drinking Fountains and Water Coolers" and covers such topics as the minimum number, spout height, spout location, controls, and clearances. Section 4.17 deals with "Toilet Stalls," the size and arrangement, water closets (height, placement, location of toilet paper dispensers, flush controls), toe clearance, doors, and grab bars.

Other facilities explicitly mentioned (by section number) in the ADAAG [11] include:

5. Restaurants and cafeterias
6. Medical care facilities
7. Business, mercantile and civic
8. Libraries
9. Accessible transient lodging
10. Transportation facilities (rail, air, boats, buses)
11. Judicial, legislative, and regulatory facilities
12. Detention and correctional facilities
13. Residential housing
14. Public rights-of-way
15. Recreation facilities

ADA and Web accessibility

The ADA also mandates that Americans of all capabilities can access government programs and services [12]. Services include information sources. In the past, the need for accessible information was met with Braille texts, large print, captioning, and other types of aids. With the introduction of the World Wide Web and other telecommunications-related information sources, however, many businesses and communities are falling short of meeting their ADA obligations for accessibility [13].

Under the ADA, there are no Access Board guidelines for Web accessibility. However, the World Wide Web Consortium (W3C [14]) has established the Web Accessibility Initiative (W3C/WAI [15]), which has in turn created and published Web accessibility guidelines [16]. The extent to which Web accessibility is covered under Titles II and III of the ADA is still being worked out in the courts [17]. Organizations such as colleges and universities that are requiring students to use the Internet to register, turn in homework assignments, etc., are subject not only to ADA requirements for accessibility, but also other federal laws beyond the scope of this document [18].

The 1992 revision and reinstatement of the Rehabilitation Act supports the ADA regarding information accessibility; it ensures that "individuals with disabilities [are able to] produce information and data, and have access to information and data, comparable to the information and data, and access, respectively, of individuals who are not individuals with disabilities" [19].

Title IV

Title IV of the ADA covers telecommunications access for people with hearing and speech disabilities [20]. Title IV requires telephone companies to establish interstate and intrastate telecommunications relay services (TRS) 24 hours a day, 7 days a week. TRS enables callers with hearing and speech disabilities who use text telephone (teletypewriter [TTY] or Telecommunications Device for the Deaf [TDDs]), and callers who use voice telephones, to communicate with each other through a third-party communications assistant. Title IV also requires closed captioning of federally funded public service announcements.

The design implications for Title IV compliance fall onto the common carriers (telephone companies) to ensure that their equipment (hardware and software) and operating procedures support TRS and the TTY/TDD signal protocols. Section 255 of the Telecommunications Act of 1996 mandates much more than Title IV in terms of accessibility requirements and designing telecommunication products and services.

Telecommunication: Section 255, Telecommunications Act of 1996

Section 255 of the Telecommunications Act reinforced the right of a person to information access with its focus on telephony. Section 255 stipulates that a manufacturer of telecommunications equipment or customer premises equipment "shall ensure that the equipment is designed, developed, and fabricated to be accessible to and usable by individuals with disabilities, if readily achievable." Telecommunications services (customer services such as billing, trouble reports, information) shall also be accessible to and usable by individuals with disabilities, if readily achievable [19]. The law goes on to state that if rendering the equipment or service accessible is not readily achievable, "a manufacturer or provider shall ensure that the equipment or service is compatible with existing peripheral devices or specialized customer premises equipment commonly used by individuals with disabilities to achieve access, if readily achievable" [19].

Telecommunication products and services have become an essential element of America's infrastructure. Designing telecommunication products and services are core areas in modern engineering educational programs. Note that Section 255 explicitly mentions accessibility and usability. Accessibility issues are not typically covered in any telecommunications or communications courses. Usability concerns are generally covered in courses on human factors, man-machine interfaces, and industrial design, but usability is again typically from the non-disabled point of view.

Section 255 also mandates that accessibility and usability design concepts and principles be applied throughout product design and fabrication. This is a key element from industry's perspective. The law and guidelines are specifying a design approach that mandates applying both accessibility and usability principles. Engineers, designers, and business administrators need to be aware of these guidelines and their implementation. The following section is unique because it covers product design, development, and evaluation.

§ 1193.23 Product design, development and evaluation

This section requires manufacturers to evaluate the accessibility, usability, and compatibility of telecommunications equipment and customer premises equipment and incorporate such evaluation throughout product design, development, and fabrication, as early and consistently as possible. Manufacturers must develop a process to ensure that products are designed, developed and fabricated to be accessible whenever it is readily achievable. Since what is readily achievable will vary according to the stage of development (i.e., some things will be readily achievable in the design phase which may not be in later phases), barriers to accessibility and usability must be identified throughout product design and development, from conceptualization to production. Moreover, usability can be seriously affected even after production, if information is not provided in an effective manner [21].

This section specifically mandates evaluating accessibility and usability across the entire design and manufacturing process. Recall the discussion of the "thought process" in design. This law is mandating that the design "thought process" explicitly includes accessibility and usability concerns; that is, these must be explicitly considered when developing the connection between "problem and possibility." As noted previously, accessibility is specified by guidelines and standards. Accessibility is a matter of knowing and following the guidelines and standards. Usability is another matter. Usability is much more subtle and subjective. Combining accessibility and usability is, in essence, specifying a universal design approach without calling it universal design. This section of the guidelines mandates an ongoing evaluation of the quality of the design; this requirement is truly unique.

The following provides an example of how detailed the guidelines are. This example deals with the control and mechanical functions associated with telecommunication devices. The entire section is presented in an effort to show the law's level of detail and introduce some of the ergonomic, perceptual, and cognitive issues that are covered by the universal design principles and associated design strategies.

§ 1193.41 Input, control, and mechanical functions

Input, control, and mechanical functions shall be locatable, identifiable, and operable in accordance with each of the following, assessed independently:

(a) Operable without vision. Provide at least one mode that does not require user vision.

(b) Operable with low vision and limited or no hearing. Provide at least one mode that permits operation by users with visual acuity between 20/70 and 20/200, without relying on audio output.

(c) Operable with little or no color perception. Provide at least one mode that does not require user color perception.

(d) Operable without hearing. Provide at least one mode that does not require user auditory perception.

(e) Operable with limited manual dexterity. Provide at least one mode that does not require user fine motor control or simultaneous actions.

(f) Operable with limited reach and strength. Provide at least one mode that is operable with user limited reach and strength.

(g) Operable without time-dependent controls. Provide at least one mode that does not require a response time. Alternatively, a response time may be required if it can be by-passed or adjusted by the user over a wide range.

(h) Operable without speech. Provide at least one mode that does not require user speech.

Chapter four: Accessible design

(i) Operable with limited cognitive skills. Provide at least one mode that minimizes the cognitive, memory, language, and learning skills required of the user [21].

The details provided in the law are, in fact, design strategies. Because these are legally mandated, they represent accessible design strategies. Designers must be aware of these requirements and acquire the knowledge and skills necessary to effectively and economically address them in their designs.

Compatibility

The law also mandates "compatibility." Compatibility is an important accessible and universal design concept. The basic idea of compatibility is simple: designed entities should not render appropriate and associated assistive technology inoperable and thereby raise barriers for individuals with impairments. The following section is unique in that it defines compatibility with respect to telecommunication products and services.

§ 1193.51 Compatibility

When required by Subpart B of this part, telecommunications equipment and customer premises equipment shall be compatible with peripheral devices and specialized customer premises equipment commonly used by individuals with disabilities to achieve accessibility, and shall comply with the following provisions, as applicable:

(a) External electronic access to all information and control mechanisms. Information needed for the operation of products (including output, alerts, icons, online help, and documentation) shall be available in a standard electronic text format on a cross-industry standard port and all input to and control of a product shall allow for real time operation by electronic text input into a cross-industry standard external port and in cross-industry standard format. The cross-industry standard port shall not require manipulation of a connector by the user.

(b) Connection point for external audio processing devices. Products providing auditory output shall provide the auditory signal at a standard signal level through an industry standard connector.

(c) Compatibility of controls with prosthetics. Touchscreen and touch-operated controls shall be operable without requiring body contact or close body proximity.

(d) TTY connectability. Products which provide a function allowing voice communication and which do not themselves provide a TTY functionality shall provide a standard non-acoustic connection point for TTYs. It shall also be possible for the user to easily turn any microphone on and off to allow the user to intermix speech with TTY use.

(e) TTY signal compatibility. Products, including those providing voice communication functionality, shall support use of all cross-manufacturer non-proprietary standard signals used by TTYs [21].

This example raises many design concerns. Most fundamentally, it requires that both telecommunication product and service designers and assistive technology designers be knowledgeable of what the other is doing. Neither group of designers deliberately wants to create undue hardships for the other group. However, serious debates ensue as to design responsibilities and incurred costs.

Digital cell phone and hearing aid compatibility provides a most relevant example. Section 255 requires compatibility: "new products shall not cause interference to hearing technologies (including hearing aids, cochlear implants, and assistive listening devices) of users or bystanders" [21]. On July 10, 2003, the Federal Communications Commission (FCC) modified the exemption for wireless phones under the Hearing Aid Compatibility Act of 1988. This means that wireless phone manufacturers and service providers must make digital wireless phones accessible to individuals who use hearing aids [22, 23]. Digital encoding and transmission of audio signals with very high carrier frequencies cause many hearing aids to buzz due to these RF transmissions [24–26]. A Motorola website on hearing aid interference provides a good description of the problem and also a collection of audio clips illustrating various forms of interference [27].

Consumer groups petitioned the FCC (Federal Communications Commission), stating the problem and asking the FCC to deal with this problem so that people with hearing aids would have the same ability to use the new digital technologies as everyone else [25]. In response, the FCC, in collaboration with consumers, telephone companies, and hearing aid manufacturers, contacted the American National Standards Institute (ANSI) to develop a measurement standard for hearing aid cell phone compatibility [25]. ANSI is a private, non-profit organization that administers and coordinates the U.S. voluntary standardization and conformity assessment system [28]. ANSI standards cover virtually every product and service. In January 2001, the ANSI C63.19 measurement standards were approved by ANSI [25].

While the ANSI C63.19 measurement standards lay out a standardized method for assessing hearing aid/cell phone compatibility, they do not address how this compatibility is to be achieved. In fact, the ANSI report recognizes the tension that exists between the cell phone manufacturers and the hearing aids manufacturers. Each side would like the other to design a product that deals with the interference problem [29]. Solutions to the interference problems are emerging in the form of new cell phone antenna designs [30] and external devices such as the Motorola Hands-Free Neckloop accessory [31].

Engineers designing telecommunication products and services should be aware of these mandates and guidelines to ensure accessibility. While companies designing, manufacturing, and marketing telecommunications products and services are knowledgeable of these laws, mandates, guidelines, and standards, very few — if any — engineering students are even aware of the existence of the Section 255, Telecommunications Act of 1996 or the Access Board. This represents a serious gap in most engineering curricula.

Electronic and information technology: Section 508 of the Rehabilitation Act Amendments of 1998

Section 508 specifies accessibility requirements for federal departments and agencies that use electronic and information technology. Electronic and information technology (E&IT) is defined as electronic technology that is used in carrying out information activities, involving any form of information [32].

The following states Section 508's mandates:

> Section 508 requires the Access Board to publish standards setting forth a definition of electronic and information technology and the technical and functional performance criteria necessary for such technology to comply with Section 508. Section 508 requires that when Federal agencies develop, procure, maintain, or use electronic and information technology, they shall ensure that the electronic and information technology allows Federal employees with disabilities to have access to and use of information and data that is comparable to the access to and use of information and data by Federal employees who are not individuals with disabilities, unless an undue burden would be imposed on the agency. Section 508 also requires that individuals with disabilities, who are members of the public seeking information or services from a Federal agency, have access to and use of information and data that is comparable to that provided to the public who are not individuals with disabilities, unless an undue burden would be imposed on the agency [33].

The implications of Section 508 for businesses are significant. Anyone designing E&IT products or services with the intention or hope of selling to federal agencies must conform to the accessibility mandates and guidelines of Section 508. In 1960, California was the first state to impose stringent environmental emission control standards for automobiles. Automakers responded by designing and building a "California car" that met the unique environmental emissions requirements solely for California [34]. E&IT

companies cannot economically build a "Federal Government" E&IT product. The federal government is simply too big a customer. Hence, the impact of Section 508 is being felt immediately and throughout all E&IT market segments.

It is understood from Section 508 that electronic and information technology addresses a broader spectrum than Information Technology alone, and includes the full breadth of the future information environment. Section 508's intent is to ensure that government employees and the public have access to the government's information environment as it evolves. As specified, information activities include, but are not limited to, "the creation, translation, duplication, serving, acquisition, manipulation, storage, management, movement, control, display, switching, interchange, transmission, or reception of data or information" [32]. Furthermore, the evolving regulations would require that the documentation (instructions, service, etc.) associated with E&IT also be accessible and useable [32]. Finally, the E&IT should not interfere with the assistive technology used daily by people with disabilities [32].

The table of contents from the final regulations "Part 1194—Electronic and Information Technology Accessibility Standards" [33] shows the breadth and scope of the law, from software, to Web-based applications, to information and documentation support. In addition, a section of the law mandates specific "Functional Performance Criteria." The law clearly defines E&IT in the broadest terms.

The following shows the major subparts of Section 508. It lays out technical standards for E&IT [33]. Subpart C specifies functional performance criteria. Subpart D provides accessibility guidelines for information, documentation, and support functions.

  Subpart B—Technical Standards:
    1194.21 Software applications and operating systems
    1194.22 Web-based intranet and internet information and applications
    1194.23 Telecommunications products
    1194.24 Video and multimedia products
    1194.25 Self contained, closed products
    1194.26 Desktop and portable computers
  Subpart C—Functional Performance Criteria:
    1194.31 Functional performance criteria
  Subpart D—Information, Documentation, and Support:
    1194.41 Information, documentation, and support

With the exception of Subpart C, Functional Performance Criteria, the accessibility standards follow from a social model and specify accessible design methods. These standards may embody universal design elements but they are basically intended to address the needs of people with disabilities and not improve performance for all people.

The subpart of Section 508 dealing with functional performance criteria (§ 1194.31) is unique in that it mandates functional performance criteria with respect to the accessibility of E&IT. This section is presented in full because it lays out accessible design criteria and provides a context for the universal design principles and design strategies to be presented in Section III.

Subpart C — Functional performance criteria

§ 1194.31 Functional performance criteria

(a) At least one mode of operation and information retrieval that does not require user vision shall be provided, or support for assistive technology used by people who are blind or visually impaired shall be provided.

(b) At least one mode of operation and information retrieval that does not require visual acuity greater than 20/70 shall be provided in audio and enlarged print output working together or independently, or support for assistive technology used by people who are visually impaired shall be provided.

(c) At least one mode of operation and information retrieval that does not require user hearing shall be provided, or support for assistive technology used by people who are deaf or hard of hearing shall be provided.

(d) Where audio information is important for the use of a product, at least one mode of operation and information retrieval shall be provided in an enhanced auditory fashion, or support for assistive hearing devices shall be provided.

(e) At least one mode of operation and information retrieval that does not require user speech shall be provided, or support for assistive technology used by people with disabilities shall be provided.

(f) At least one mode of operation and information retrieval that does not require fine motor control or simultaneous actions and that is operable with limited reach and strength shall be provided [33].

These criteria represent a major shift in legal language, and follow from a systems model and embody key elements of universal design. The criteria are in essence mandating a multimedia approach to design, an approach that is multi-sensory in terms of both communicating with and controlling an E&IT entity. Adherence to these functional performance criteria are in essence adherence to some of the fundamental principles of universal design.

In summary, Section 508 is a complex law covering accessibility for all current and future E&IT products and services purchased by the federal government. The law has language reflecting both the social and systems conceptual models of disability. The law and associated guidelines and standards cover accessible design methods and strategies for E&IT entities. The functional performance criteria subpart deviates from the language of the remaining subsections and takes a much more holistic approach to accessibility. Such an approach exemplifies universal design.

The impact of Section 508 is huge. Its influence is and will continue to ripple through all segments of American society. The federal government is such a large purchaser of E&IT that this law is, in practice, expanding the accessibility of E&IT for everyone and in so doing is helping to redefine society's conception of disability. Section 508's accessibility mandates for E&IT are precisely the enabling environmental factors, discussed in the WHO-ICF model of disability. The evolving E&IT will change what it means to be disabled in jobs and other activities that utilize E&IT.

Access Board accessible design technical assistance

The Access Board provides technical assistance for implementing the accessibility guidelines [35]. The technical assistance includes online tutorials, help in producing Web-based tutorials, and design assistance for each of the guidelines. The topical organization for the technical support documents follows a standardized format [36]. Table 4.2 illustrates the standardized format.

Table 4.3 considers the Accessible Telecommunications Product Design Tutorial, which covers § 1194.23, Telecommunications Products [36].

This example shows the structure and level of detail of the online technical assistance tutorial. Such support is available for all the Access Board guidelines and standards [37]. These guidelines and technical assistance tutorials provide the accessible design strategies that a designer can apply to satisfy a specific accessibility requirement. It is important that designers are aware of these technology assistance materials because they can save a lot of time, energy, resources, and ultimately money.

Accessibility forum

Section 508 states that any E&IT provider to the federal government must provide products and services that are in compliance with Section 508 mandates. The Government Service Administration (GSA) is responsible for federal government purchases and must ensure that the products and services are in compliance with Section 508. The Accessibility Forum was created to address the mutual concerns of buyers and sellers. The following is taken from the Accessibility Forum website.

> The Accessibility Forum began in May of 2001 as an ongoing collaboration among stakeholders affected by Section 508 including user, industry, government, and other communities. The Forum membership includes over 640 organizations from industry (electronics, information technology, and assistive technology),

Chapter four: Accessible design 55

Table 4.2 Standardized Format for the Access Board's Design Assistance Documents

Topical area	Description of content
1. Provision	Quotes the provision verbatim from the sections. The short headings for the provisions are taken from the actual text of the provisions and are consistent with the headings used in previously developed Access Board technical assistance guides.
2. Introduction and Background	Gives an introduction to the provision and the issue being addressed by the rule.
3. Design Guidance	Uses a question-and-answer format, in most provisions, to address the interests of product designers. This section provides further technical information and gives measurable targets. Where applicable, examples of implementation methods are given. It is imperative that federal agencies understand that these methods illustrated are not the only approaches for reaching accessibility goals.
4. Requirements and Recommendations	Wraps-up discussion of the provision. This last section helps identify the requirements of the provision. In addition, it gives recommendations on ways that fuller accessibility may be provided: *Provision Requirements* are design elements necessary for product conformance to the provision. *Recommended Practices* are design approaches that augment or exceed the provision requirement and enhance usability and accessibility.

Source: See Access Board, Technical Assistance, Training, and Research, http://www.access-board.gov/indexes/technicalindex.htm, [36].

associations for people with disabilities, the research community (academia, research institutions, and standards groups), and government agencies. The General Services Administration (GSA) has been the initial sponsor for the Forum.

The Accessibility Forum focuses on long-term solutions. Projects were defined to produce results that assist government in making informed decisions about Section 508 related procurement. The Forum completed two projects for research on two high priority sets of issues—Objective measures for assessing compliance with the Access Board standard and AT/E&IT Interoperability between assistive technology products and mainline information technology products. The Forum also provides a means for government, industry, and users to communicate on issues and areas where further effort could enhance E&IT accessibility.

GSA established a Forum Staff to carry on day-to-day Forum efforts. The staff is composed of government personnel, including

Table 4.3 Standardized Format for the Access Board's Design Assistance Documents[a]

Topical area	Description of content
508 Provision § 1194.23(f)	For transmitted voice signals, telecommunications products shall provide a gain adjustable up to a minimum of 20 dB. For incremental volume control, at least one intermediate step of 12 dB of gain shall be provided.
Introduction and Background	People with hearing loss generally require additional volume to hear effectively. A user may not be able to understand speech at nominal volume levels. This provision enables people who are hard of hearing to increase their telephone volume gain in order to maximize their residual hearing.
	Many people with hearing loss are more sensitive to noise and less able to recognize words in the presence of noise than individuals with normal hearing. Therefore, increased volume assists these users in achieving a volume level and speech-to-noise ratio sufficient for their needs.
Design Guidance	• Is the company product affected by this provision?
	This provision applies to all telecommunications products that transmit a two-way voice signal. Products that only provide one-way communications, such as speakers that give operating instructions to the user, are not covered by this provision.
	• dB is a relative term, what is the reference quantity?
	The term dB is a logarithmic unit used to describe a ratio. When used with SPL (e.g., 65 db SPL) it expresses an absolute measure of sound pressure level (against no sound). Most telecommunications standards now define gain in terms of receive output loudness rating, not sound pressure level.
	• What gain level is required?
	This provision requires products to be equipped with volume control that provides amplification adjustable to a gain of at least 20 dB above the default volume. If a volume adjustment is provided that allows a user to set the gain anywhere from 0 to the minimum requirement of 20 dB gain above default, then there is no need to specify an intermediate step of 12 dB. If a stepped volume control is provided, one of the intermediate levels must provide 12 dB of gain above default. Some telephones are set with a higher default than others. These telephones are not given "credit" for providing a high default setting and must still provide 20 dB of gain above their default setting in order to conform to this provision. Some phones may allow the user to reset the default volume. In these cases the default is that setting that the manufacturer normally uses for the telephones as sales time.

Table 4.3 Standardized Format for the Access Board's Design Assistance Documents[a] (Continued)

Topical area	Description of content
Provision Requirements	• The acoustic output should be controllable by the user. • For incremental volume control, at least one intermediate step of 12 dB of gain should be provided. • The maximum volume setting should reach at least 20 dB above the default volume setting.
Recommended Practices	The intention of this requirement is that a usable signal be available at the maximum volume setting. Commonly accepted frequency responses and distortion/clipping levels could be accomplished within acceptable range at maximum volume setting.

[a] This table covers *Volume Control* for telecommunication products.

Source: See Access Board, Technical Assistance, Training, and Research, http://www.access-board.gov/indexes/technicalindex.htm, [36].

an experienced Program Director and other supporting personnel, and a team of contracted organizations. The membership is moving towards an electronic based Forum and provides support in research, Web-based meeting coordination and planning, publicity activities, and all other membership efforts [38].

To help the GSA in its purchasing, and to help E&IT providers understand which E&IT features are covered by Section 508, the Accessibility Forum has created the *Section 508 Buy Wizard* [39]. The Wizard guides a user through a series of questions and information gathering stages that helps both the GSA buyer and the E&IT providers understand which, if any, specific sections of Section 508 apply to a particular E&IT product.

As noted, the Accessibility Forum has more than 640 members. It is an impressive list, and its size demonstrates the importance of Section 508 and its implications for the design of E&IT products and services, and the need to train engineers and designers in accessible and universal design principles. Member organizations include Adobe Systems, Inc.; America Online, Inc.; Bell South; Booz Allen & Hamilton; Cingular Wireless LLC; Dell Computer Corporation; Diebold; Don Johnston; Ford Motor Company; Gateway Computer; Hewlett Packard; IBM; IEEE Computer Society; Lockheed Martin; Motorola; Northrop Grumman IT; Panasonic; Sony Electronics; Sun Microsystems; Tash Inc.; TRW Systems; Underwriters Laboratories Inc.; U.S. Access Board; Xerox Corporation; and a large number of U.S. government agencies and departments.

Conclusions

This chapter summarized the major laws mandating accessible design in the United States. The legal initiatives in the United States are part of a global movement toward a systems model of disability as exemplified by the WHO's new ICF [40] and expressed in the United Nations document entitled "The Standard Rules on the Equalization of Opportunities for Persons with Disabilities" [41].

The Standard Rules consist of 22 rules concerning disabled persons. The rules fall into four chapters: (1) preconditions for equal participation, (2) target areas for equal participation, (3) implementation measures, and (4) the monitoring mechanism. These chapters cover all aspects of life of disabled persons [41]. "The Standard Rules represent a strong moral and political commitment of governments to take action to attain equalization of opportunities for persons with disabilities. The rules serve as an instrument for policy-making and as a basis for technical and economic cooperation" [39].

Virtually every developed country around the world has established laws similar to those found in the United States [42, 43]. A report by Kennig and Ryhl, entitled "Teaching Universal Design: Global Examples of Projects and Models for Teaching in Universal Design at Schools of Design and Architecture" [44], summarizes the numerous and varied global initiatives and projects directed at the teaching of universal design principles. The Accessibility Forum has a resource website that provides links to sites detailing international accessibility efforts [43]. Hence, the legal initiatives and support structures such as the Access Board and Accessibility Forum are U.S. expressions of a broader, global movement.

From a designer's perspective, the resources provided by the Access Board are substantial and in a very practical way lay down the essential principles of accessible design. Engineers and designers should be aware of the resources available from the Access Board and take advantage of these resources in their own design activities.

The Accessibility Forum is another invaluable resource to any company doing business with the U.S. government. The *Section 508 Buy Wizard* [39] is just one example of the resources provided. Another example is the Accessibility Forum's resource site [45].

Universal and accessible design is evolving in a global context. Global market conditions include a worldwide consumer base wherein consumers have a large number of options and can easily and quickly change suppliers, many suppliers are spread out across the globe, and the suppliers are vulnerable in that they are under pressure for faster product innovation and design turn-around time with a very dynamic market [46]. The global initiatives to address the needs and civil rights of the world's disabled people is another driving force [39], and the changing global conceptualization of disability is exemplified by the WHO-ICF [40]. All of these elements combine to create an urgency with respect to understanding and being able to apply universal and accessible design principles. The next section (Section III) presents the universal design principles, along with specific design strategies and examples.

References

1. Access Board, "About the U.S. Access Board," http://www.access-board.gov/about.htm, 2006.
2. Access Board, "Members of the Board," http://www.access-board.gov/about/members.htm, 2006.
3. Access Board, "Technical Assistance, Training, and Research," http://www.access-board.gov/training.htm, 2007.
4. Access Board, Training — provides information on Access Board training programs, http://www.access-board.gov/training.htm, 2004.
5. Access Board, Uniform Federal Accessibility Standards (UFAS). Washington, D.C.: Federal Register, 49 FR 31528, 1984.
6. Access Board, "Americans with Disabilities Act Accessibility Requirements," http://www.access-board.gov/bfdg/adares.htm., 1999.
7. R. F. Erlandson, M. J. Noblet, and J. A. Phelps, "Impact of Poka-Yoke Device on Job Performance of Individuals with Cognitive Impairments," *IEEE Trans. Rehab. Eng.*, 6, 269–276, 1998.
8. R. F. Erlandson and D. Sant, "Poka-Yoke Process Controller Designed for Individuals with Cognitive Impairments," *Assistive Technol.*, 10, 102–112, 1998.
9. R. F. Erlandson, "Accessible Design and Employment of People with Disabilities," in *The Sourcebook of Rehabilitation and Mental Health Practice*, D. Moxley and J. Finch, Eds. New York: Plenum, 2003, pp. 235–252.
10. Access Board, "Chart: U.S. Architectural and Transportation Barriers Compliance Board Americans With Disabilities Act," http://www.access-board.gov/publications/ADAFactSheet/A13.html, 1992.
11. Access Board, "ADA Accessibility Guidelines for Buildings and Facilities (ADAAG)," http://www.access-board.gov/adaag/html/adaag.htm, 2002.
12. Access Board, "Revised ADA and ABA Accessibility Guidelines. Published in the Federal Register July 23, 2004 and amended August 5, 2005," http://www.access-board.gov/ada-aba/final.htm, 2005.
13. C. D. Waddell, *The Growing Digital Divide in Access for People with Disabilities: Overcoming Barriers to Participation*. San Jose, CA: Office of Equity Assurance, 1999.
14. W3C, World Wide Web Consortium-Homepage, http://www.w3.org/, 2002.
15. WAI, "About WAI," http://www.w3.org/WAI/about.html, 2002.
16. WAI/W3C, "Web Content Accessibility Guidelines 1.0," http://www.w3.org/TR/1999/WAI-WEBCONTENT-19990505/, 1999.
17. C. D. Waddell, "Applying the ADA to the Internet: A Web Accessibility Standard," http://www.rit.edu/~easi/law/weblaw1.htm, 1998.
18. M. D. Cohen, "Comparative Analysis: IDEA, Section 504 and the ADA," http://at-advocacy.phillynews.com/misc/cohen2.html, 1998.
19. FCC, "Access to Telecommunications Service, Telecommunications Equipment and Customer Premises Equipment by Persons with Disabilities," *Federal Register*, 64, 63235–63258, 1999.

20. FCC, "Title IV of the Americans with Disabilities Act (ADA) — Telecommunications services for hearing-impaired and speech-impaired individuals codified at 47 U.S.C. § 225," http://www.fcc.gov/cgb/dro/title4.htm, 2003.
21. Access Board, *Telecommunications Act Accessibility Guidelines: Final Rule*. Washington, D.C.: Federal Register, 36 CFR Part 1193, 1998.
22. Technology Access Program, "Digital Wireless Telephones and Hearing Aids: An Introduction," http://tap.gallaudet.edu/digwirelessintro.htm, 2002.
23. M. Ross, "Wireless Telephones and Hearing Aids: An Overview," *J. Am. Acad. Audiol.*, 12, 286–289, 2001.
24. D. D. Hoolihan, "ANSI C63.19: Establishing Compatibility between Hearing Aids and Cellular Telephones," http://www.ce-mag.com/archive/01/spring/Hoolihan.html, 2001.
25. H. S. Berger, "Hearing Aid / Cellular Telephone Compatibility: ANSI C63.19," ANSI 2001.
26. H. Levitt and J. Harkins, "Digital Wireless Telephones and Hearing Aids: New Challenges for Audiology," *J. Am. Acad. Audiol.*, 12, 273–274, 2001.
27. Motorola Inc., "Hearing Aid Interference," ,http://commerce.motorola.com/consumer/QWhtml/accessibility/hearingAid.html, 2003.
28. ANSI, ANSI: American National Standards Institute Home Page, http://www.ansi.org/, 2005.
29. ANSI, "ANSI C63.19: American National Standard for Methods of Measurement of Compatibility between Wireless Communication Devices and Hearing Aids," ANSI 2001.
30. RFSafe.Com, "RF Safe Approved Cell Phones Are ADA & FCC Compliant Already." http://www.rfsafe.com/an_rf_safe_approved_phone_is_ada.htm, 2004.
31. Motorola Inc., "Hands-Free Neckloop accessory," http://commerce.motorola.com/consumer/QWhtml/accessibility/prodservices.html, 2004.
32. Electronic and Information Technology Access Advisory Committee, "Final Report of the EITAAC," http://www.access-board.gov/pubs/eitaacrpt.htm, 1999.
33. Access Board, Electronic and Information Technology Accessibility Standards. Washington, D.C.: Federal Register, 36 CFR Part 1194, 2000.
34. SAE International, History of SAE and Automotive Industry (1905–2002), http://www.sae.org/sae100/history/timeline/timeline.htm, 2004.
35. Access Board, Publications of the Access Board: Guidelines, training and technical support material, research, and enforcement information. This is a valuable resource site for designers, planners and managers. http://www.access-board.gov/pubs.htm, 2006.
36. Access Board, "Accessible Telecommunications Product Design Technical Assistance," http://www.access-board.gov/sec508/telecom-course.htm#Volume%20Control, 2001.
37. Access Board, "Annotated Bibliography and Links to the Resources, "http://www.access-board.gov/sec508/refresh/reading.htm, 2007.
38. Accessibility Forum, "About the Accessibility Forum," http://www.accessibilityforum.org/about_forum/index.html, 2006.
39. Accessibility Forum, "Buy Accessible Wizard," http://www.buyaccessible.gov/, 2004.
40. WHO, "Towards a Common Language for Functioning, Disability and Health: ICF, World Health Organization, Geneva, 2002.

41. United Nations Department of Economic and Social Affairs, "The Standard Rules on the Equalization of Opportunities for Persons with Disabilities," http://www.un.org/esa/socdev/enable/dissre00.htm, adopted by the United Nations General Assembly, 48th session, Resolution 48/96, Annex, of 20 December 1993.
42. W. F. E. Preiser and E. Ostroff, *Universal Design Handbook*, New York: McGraw-Hill, 2001.
43. Accessibility Forum, "Online Resources/Related Sites: International Sites," 2004: Accessibility Forum, 2004.
44. B. Kennig and C. Ryhl, "Teaching Universal Design: Global Examples of Projects and Models for Teaching in Universal Design at Schools of Design and Architecture," AAOutils: ANLH, Brussels, 2002.
45. Accessibility Forum, "Online Resources/Related Sites: International Sites," http://www.accessibilityforum.org/resources/international.html, 2006.
46. R. Reich, *The Future of Success*. New York: Alfred A. Knopf, 2000.

section three

Universal design principles, strategies, and examples

Section goals:

- Present the hierarchical structure of the principles and the implications of this structure on the design process.
- Present the principles of universal design.
- Provide a rationale for each principle so that designers will be inclined to utilize the principles as appropriate.
- Provide design strategies and examples for each universal design principle.
- Explore the use of universal design principles in the workplace.
- Show how universal and accessible design have been integrated into the Web design community.

Scope of the discussion

This section focuses on universal design. As noted in Section II, there is an overlap between universal and accessible design. Hence, where universal design principles and strategies increase accessibility, they can be applied to both universal and accessible design.

This section divides the universal design principles into three categories: (1) person focused, (2) process focused, and (3) transcending. Specific design strategies are presented and discussed for each principle. Examples will illustrate applications of the various strategies. Arguably, more design strategies could be presented. However, the strategies presented herein provide a solid base from which to proceed implementing more universally designed and accessible products and processes.

A collection of design strategies is presented for each universal design principle. The term "design strategy" means concrete, pragmatic ways to implement and build entities so that they, in fact, possess the properties and characteristics espoused by the universal design principles. The examples have been chosen to span as many engineering and design disciplines as possible. The strategies derive from human safety considerations, differences in human size and shape, native abilities, and the common elements of basic human functioning. The examples emphasize that the pressures for universal design derive not only from laws and regulations, but also from strong global market pressures. The universal design principles work best when applied in a systematic and coherent manner because the combined effects are synergistic. Such synergisms are noted and highlighted as appropriate.

Universal design strategies derive from many sources and as such apply to the entire spectrum of human design activity. The umbrella concept of universal design brings these existing design strategies together in new ways, and that is the uniqueness and strength of the universal design approach. While the origins of universal and accessible design are in architecture, civil engineering, and related disciplines, the concepts have spread to virtually all human design endeavors. To design entities that will be in conformance with the definition of universal design, the entities should be:

1. Ergonomically sound
2. Perceptible
3. Cognitively sound
4. Flexible
5. Error-managed (proofed)
6. Efficient
7. Stable and predictable
8. Equitable

The principles of universal design can be divided into three broad categories: (1) those dealing primarily with the person, (2) those dealing primarily with process, and (3) those that transcend both people and process. The hierarchical structure of the universal design principles and the implications of this hierarchy for the design process are presented in Chapter 5. Three principles—ergonomically sound, perceptible, and cognitively sound—can be termed *human factors principles* in that they focus on the person. These three principles are discussed in Chapters 6, 7, and 8, respectively.

The principles dealing with flexibility, error-management, efficiency, and stability and predictability can be termed *process principles* in that they focus on broader process concerns. These principles are the focus of Chapters 9 through 12, respectively. The principle of equitability, discussed in Chapter 13, integrates and transcends the other principles. The principle of equitability places constraints on how all the other principles are applied.

The two final chapters in this section step back and consider universal design applications from broader perspectives. Chapter 14 discusses the use of universal design principles in the workplace. The work environment is sufficiently complex and unique to warrant a closer examination and analysis. Chapter 15 focuses on Web accessibility and Web design. This is a separate topic for two reasons: (1) the ubiquity of Web-based activity and its global impact, and (2) Web design exemplifies the mainstreaming of

Section three: Universal design principles, strategies, and examples 65

accessible and universal design principles into the design process of a design community and as such provides a powerful way to summarize the material in Sections II and III.

chapter five

Hierarchical structure of universal design principles and implications for the design process

Chapter goals:

- Present and explain the hierarchical structure of universal design principles.
- Discuss how this hierarchical structure affects the design process.

Design philosophy

Universal and accessible design strategies derive from a simple fundamental premise: design entities to work with, support, and enhance human functioning. Design to human strengths and not human weaknesses. Create products and processes wherein people are allowed to do what they do best, and where the products and processes are required to do what they do best. It is from this perspective that universal design is defined as the design of entities that can be used and experienced by people of all abilities, to the greatest extent possible, without adaptations.

Universal design principles

This book presents the following eight areas of focus and design principles:

1. Ergonomically sound
2. Perceptible
3. Cognitively sound
4. Flexible
5. Error-managed (proofed)
6. Efficient
7. Stable and predictable
8. Equitable

```
Higher level          Transcending principles      More general
places design                   |                  More encompassing
constraints on the              |
lower level           Process related principles
                                |
                                |                  More detailed
                      Human factors principles     More narrowly defined
                                                   More specific
```

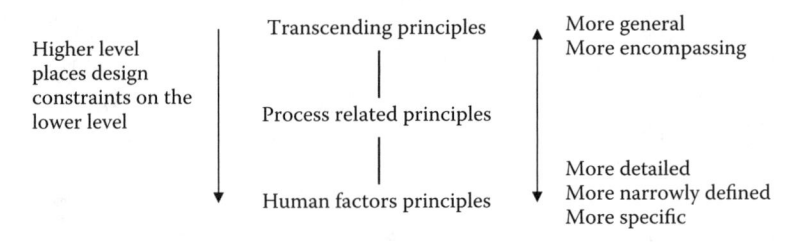

Figure 5.1 The hierarchical structure of the universal design principles. The transcending principles are value based and place constraints on the use of process principles, which in turn place constraints on the application of the human factors principles.

These eight principles can be divided into three major categories, depending on whether they primarily address person-centered concerns, process concerns, or transcendental concerns. The principles dealing with ergonomics, perception, and cognition are termed the human factors principles. The principles dealing with, flexibility, error-management, efficiency, and stability/predictability are termed process principles because they deal with activities and participation (that is, processes). The final principle deals with equity and as such is very different from the others. Equity is a value judgment. As a design community, we are stating that we desire universally designed entities to be equitable.

The three categories of universal design principles also have an implicit order associated with them that can be used to characterize the universal design principle categories as a hierarchical structure. Figure 5.1 illustrates this structure. Each of these categories will now be considered in more detail.

Principles targeted at the person (human factors)

The principles aimed primarily at the person deal with ergonomics, perception, and cognitive concerns. These issues focus on human factors and are centered in the person. The person must be able to turn the knob, see the signal, hear a prompt, and understand the operating instructions. The design process must make sure the person can access and use an entity.

The traditional field of ergonomics is a rich source of methods and strategies [1–5]. The International Ergonomics Association (IEA) defines ergonomics (or human factors) as "the scientific discipline concerned with the understanding of interactions among humans and other elements of a system, and the profession that applies theory, principles, data, and methods to design in order to optimize human well-being and overall system performance" [6]. The IEA embraces a broad definition of ergonomics, that is, "ergonomics promotes a holistic approach in which considerations of physical, cognitive, social, organizational, environmental and other relevant factors are taken into account" [6]. According to The Ergonomics Society, "Ergonomic design is a way of considering design options to ensure that people's capabilities and limitations are taken into account. This helps to ensure that the product is fit for use by the target users" [7]. For universal design, the target users are defined as broadly as possible.

For this discussion, *ergonomics* will be viewed more traditionally and narrowly than the IEA definition. Ergonomics will refer to the physical demands of an activity, task, or job. Ergonomics, together with the universal design principles dealing with perception and cognitive elements, cover the broader *human factors* view.

Principles targeted at the process

A process can be defined as a collection of related tasks or activities that lead to a particular result. People perform tasks and participate in activities that lead to results. Processes are designed entities and exhibit characteristics that can create barriers for people or support and facilitate human performance. Those principles aimed primarily at the process deal with flexibility, error-management, efficiency, and stability or predictability.

Because processes include people, it is not surprising that the process-related design strategies impose constraints and influence the application of the human factors-based design strategies. This will be seen time and again as the process-related design strategies are presented. Also, the process-related strategies are interrelated. For example, error-proofing a process will make it more efficient as well as more stable and predictable. This interrelatedness is a source of synergism among the design principles and their respective design strategies.

Transcending, integrating principle

Universally designed entities should be equitable in that the entities should provide the same means of use for all users: identical whenever possible, equivalent when not [8]. Equitability is a value. Many products and services are designed specifically so as not to be equitable. For example, luxury items such as jewelry, with an associated brand name, are meant to distinguish the owner and not be accessible by everyone.

As a value, equitability transcends or goes beyond the other universal design principles in that if it is satisfied, it integrates elements from all the other principles and exhibits features and characteristics that are unique to being equitable. Such unique features and characteristics include being age and context appropriate, being aesthetically pleasing, being affordable, and having a broad market appeal. These design features and characteristics are generally true for any product or service, but is especially true for designers seeking to follow universal design principles.

Equitability imposes constraints on the other design principles in that they must be applied so that the designed entities are accepted by a broad spectrum of users. The designs must be age and context appropriate as well as aesthetically pleasing. In a most fundamental way, equitability forces the integration of the other universal design principles.

Accessible design does not concern itself with being equitable. Accessible design is concerned with removing or preventing environmental barriers as prescribed by law, guidelines, and standards. This does not mean that an accessible design requirement cannot be equitable.

Equitability will be the final universal design principle considered in this discussion, its representation as the synergistic integration of the other universal design principles, argues for its prominence in design requirements. This is why it is often the first-cited universal design principle in other discussions.

Hierarchical constraints

As conceived, the universal design principles form the hierarchical structure depicted diagrammatically in Figure 5.1. Michael Polanyi, a distinguished scientist, system theorist, and philosopher, in his paper entitled "Life's Irreducible Structure" [9], presents a general systems theoretic principle of such hierarchical systems. The principle states that the higher level places constraints on the structure or design of the lower level, and furthermore, one can neither express the higher level's operational laws in terms of the lower-level operational laws, nor can one express the lower-level laws in terms of the higher-level laws.

Operational laws and principles, such as the various psychometric laws, physiological principles, psychological principles, and the biochemistry of brain and neurological and neuromuscular functioning, form the basis for what and how people behave and function with respect to the human factors principles. The process principles place constraints on the various human factors principles and design strategies. For example, the universal design process principle, efficiency, seeks to reduce or remove non-value-added activity (NVAA) from processes. Lifting and carrying materials from one place to another are typically NVAAs. From a human factors (ergonomics) perspective, the job might be designed so that the lifting and carrying conform to accepted ergonomic standards. However, from process design perspective, one seeks to eliminate lifting and carrying activities in an effort to remove NVAAs. In this way, the process design places constraints on the human factors level.

Consider error-proofing as another example of this phenomenon. An error-proofing requirement might be to not allow a driver to put diesel fuel into a car requiring unleaded gasoline. The process (context) dictates what elements are available to the designer and how they can be used. In terms of the human factors principles, this error-proofing task suggests using a forcing function as a natural constraint. Specifically, make the diesel fuel nozzle bigger than the unleaded nozzle and size the gas tank receptacle accordingly. In this way, the diesel nozzle cannot fit into the nonleaded receptacle. In terms of the human factors principles, this provides a forcing function, natural constraint. More will be discussed about these strategies later; the main point of this discussion is that the universal design principles of the higher levels place design constraints on the design principles at the lower levels.

References

1. E. Grandjean, *Fitting the Task to the Man*, 4th ed. New York: Taylor & Francis, 1991.
2. G. D. Jimmeerson, C. J. Jacobs, and D. S. Fischer, "Design for Ergonomics in Manufacturing," National Center for Manufacturing Sciences, Ann Arbor, MI: White Paper, 1993.
3. Ergonomics Society Press Release, "Ergonomics in Car Design Benefits Older Drivers," http://www.ergonomics.org.uk/press/press2.htm, 1999.
4. R. Ahsan, "Ergonomics Consequences of Assistive Technology for the Disabled," presented at Proceedings of Joint Asia-Pacific Computer Human Interaction/, ASEAN Ergonomics, 2000.
5. V. Chandramohan, "Incorporating Ergonomics into an Integrated QFD/DFX Approach for Concurrent Product Design Evaluation," *Proceedings of the Joint Conference of APCHI 2000 and ASEAN Ergonomics 2000*, 2000, pp. 75–80.
6. International Ergonomics Association, "The Discipline of Ergonomics," http://www.iea.cc/ergonomics/, 2000.
7. The Ergonomics Society, "Ergonomics: Definition," http://www.ergonomics.org.uk/ergonomics.htm, 2004.
8. B. R. Connell, M. Jones, R. Mace, J. Mueller, A. Mullick, E. Ostroff, J. Sanford, E. Steinfeld, M. Story, and G. Vanderheiden, "Principles of Universal Design," http://www.design.ncsu.edu:8120/cud/univ_design/princ_overview.htm, 1997.
9. M. Polanyi, "Life's Irreducible Structure," *Science*, 1968, pp. 1308–1212.

chapter six

Ergonomically sound

Chapter goals:

- State the ergonomically sound universal design principle.
- Provide a rationale for the ergonomically sound design principle's relevance.
- Provide design strategies that address the ergonomically sound design principle.
- Provide examples of applying the design strategies.

Ergonomically sound

Principle

The physical demands associated with the use of an entity must be within acceptable limits for a wide range of users.

Discussion

Ergonomics refers to the physical demands of an activity, task, or job. Ergonomic design "is a way of considering design options to ensure that people's capabilities and limitations are taken into account. This helps to ensure that the product is fit for use by the target users" [1]. Humans share common physiological, anatomical, and biochemical attributes. These common attributes mean that people tend to respond the same way to exercise and activities in terms of fatigue, repetitive motion injuries, and general wear and tear on body parts. Our common attributes also mean that common strategies and techniques can be designed to reduce the adverse effects of work, exercise, and strenuous activity. For example, Figure 6.1 shows a person carrying a heavy load.

Personal experience tells us that it is easier to carry a heavy weight close to the body rather than extended. Physics provides the explanation in terms of the different torques generated. Case (a) generates a higher torque at the

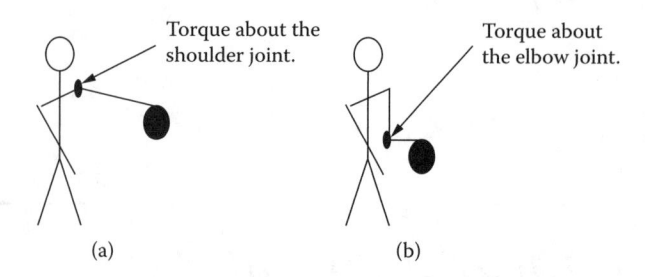

Figure 6.1 Two stick figures are shown carrying a load. In (a), the arm is extended and the elbow locked. In (b), the elbow is bent.

shoulder than the torque generated by case (b) at the elbow. Furthermore, the affected shoulder muscles are typically not as strong as the biceps muscles in the arm. Given the combination of less torque and more muscle strength, case (b) is the preferred lifting or carrying strategy.

The combination of common human physiology and anatomy, and the laws of physics allows designers to design lifting and carrying strategies that best serve individuals. If a person has a physical impairment, then the same design approach is used: look at the physiology, anatomy, and the forces (torques) generated by the load, and design a strategy that maximizes the person's capabilities while minimizing the work in terms of applied torques or forces. This is the fundamental approach to ergonomic universal design strategies. Use a person's capabilities in the most efficient and effective manner possible while avoiding injury.

There are, of course, individual variations in height, muscle mass, metabolic rates, arm length, leg length, and all the physiological, anatomical, and biochemical attributes. Gender, race, age, injury, disease, and impairments due to developmental disabilities introduce further variations and variability. Universal design strategies must address such variations and variability.

Ramps, to access buildings, exemplify universal design. People can push strollers or pull luggage up ramps. People using walkers can more easily negotiate a ramp than stairs. People of any height or weight can use ramps more easily than stairs. Wheelchair users can use ramps provided that the ramps are not so steep that wheelchair users cannot move themselves up the ramp, or if necessary be pushed up the ramp.

The ramp shown in Figure 6.2 is from a museum and therefore must conform to Americans with Disabilities Act Accessibility Guidelines (ADAAG) standards. The ADAAG standards for ramps focus on technical specifications such as rise, slope, etc. and do not address the issues of an aesthetically pleasing design. The ramp in Figure 6.2 demonstrates that designers can be in compliance with accessibility guidelines and yet have a pleasing design.

Providing adequate space and surface materials for wheelchair mobility and navigation in public spaces, buildings, and facilities is another universal

Chapter six: Ergonomically sound 75

Figure 6.2 Two views are shown of the entrance to one of the buildings at The Henry Ford's Greenfield Village. Being a museum, it must conform to the Americans with Disabilities Act Accessibility Guidelines (ADAAG) standards. Clearly accessible design can be aesthetically pleasing.

design strategy. Space large enough and suitable for wheelchair users will ensure comfortable movement for most people, including people with strollers, walkers, wheeled luggage, and large wheeled brief cases.

Consumer products should be designed so that users can easily turn knobs, move switches, and make adjustments consistent with normal operations. Such modifications exemplify universal design; however, they are mandated for telecommunication devices. Recall the example from Chapter 4 (§ 1193.41 Input, control, and mechanical functions) from the final rule of the Telecommunications Act 1996 [2]. In particular, the device should be operable with "limited manual dexterity" and "limited reach and strength" [2].

Ergonomics have long been associated with work. The general principle of universal design in the workplace is to have the physical demands of work activities within acceptable limits for a wide range of workers. All individuals must be able to physically access the activity. The relative placement of materials, equipment, and workers should be ergonomically sound. Workers should be able to easily lift items, carry objects, turn knobs, press keys, and/or move computer mice; if they cannot easily perform such ergonomic tasks, alternative means of physical engagement must be available.

In 1999, the Occupational Safety and Health Administration (OSHA) published its proposed ergonomics program standards to address work-related musculoskeletal disorders (MSDs) [3]. Work-related MSDs are currently the leading cause of lost workday injuries and workers' compensation costs [3]. The ergonomic standards are intended to reduce worker exposure to ergonomic risk factors. This material is significant in that it represents another source of legally mandated environmental and human performance guidelines and standards.

The OSHA requirements were not created to deal with enhanced accessibility for individuals with disabilities. Instead, OSHA is responsible for overseeing the health and safety of American workers. The OSHA standards represent universal design strategies because applying these standards improves job performance for all workers. If a job's physical and cognitive demands are reduced, the job naturally becomes more accessible. Reducing physical and cognitive demands of a job removes environmental barriers. Ironically, increased accessibility was an unintended outcome of OSHA's ergonomic standards.

Design strategies

The following three strategies cover the essential aspects of ergonomics with respect to the physical demands placed on the person by the designed entity. These strategies are broadly stated and make reference to well-established disciplines involved with human physical activity. Such disciplines include industrial engineering, ergonomics, physiology, occupational therapy, and physical therapy. The intent here is to explicitly state the design strategies that derive from these various sources.

Strategy 1: Design to avoid ergonomic risk factors

Ergonomic risk factors include forceful exertions, dynamic motions, repetition, awkward postures, static postures, contact stress, vibration, and excessive temperatures (hot or cold) [3]. The legal imperative for this strategy derives from the 1999 proposed ergonomics standards produced by OSHA, which is part of the U.S. Department of Labor.

The OSHA standards define *ergonomics* as the science of fitting jobs to people [3]. Ergonomists have an extensive collection of tools and techniques for identifying ergonomic risk factors and creating ergonomically sound work environments. *Engineering controls* and *administrative controls* are two broad categories of intervention techniques. The term "engineering controls," as applied to the elimination or reduction of work-related hazards, includes changing, modifying, or redesigning workstations, tools, facilities, equipment, materials, and processes [3]. The engineering controls should impact ergonomic risk factors. Administrative controls can be used to reduce the magnitude, frequency, or duration of exposure to ergonomic risk factors [3]. Administrative controls include employee rotation, job task enlargement, alternative tasks, and employer-authorized changes in the workplace (for example, working in a different part of the plant, or working at home) [3].

From a more general design perspective, it simply makes sense to design entities that minimize a person's exposure to ergonomic risk factors.

Strategy 2: Design for a wide range of body sizes and shapes

Central to a person's capabilities are body size and shape. For example, body size and shape determine a person's range-of-motion and reach. The discipline of anthropometry measures and studies the human body, especially on a comparative basis [4]. Anthropometry plays a key role in ergonomic design.

The following quote provides a personal perspective on anthropometry:

> "Living in London is no fun if you're five foot," says Dee O'Connell. "But a shopping trip to the Far East can change your self-perception forever." O'Connell goes on: "I knew this was going to turn out well when my feet reached the floor of the Underground carriage. At five foot nothing, my feet rarely touch the ground when I'm sitting down, but I was in Hong Kong, a place built entirely for people like me. ...
>
> I had my first taste of what it's like to be average in a boutique next to the market. I picked out armfuls of dresses and skirts in my usual size, small. The sales assistant took them from me and said: 'You medium.' I had never before been a medium in my life, but I took her word for it, and went into the changing room. .... The hem sat just on my knee, the belt went around my waist, rather than my backside, and it didn't gape around the armhole. It cost a grand total of £15, and I would have paid 10 times that for the sheer novelty value. Obviously, I bought it" [5].

At various times, all people think about their bodies. They may worry about weight, muscle tone, and/or stamina. Yet, as evidenced by O'Connell's quote, there are individuals who, for various reasons, are compelled to deal with a world that literally does not fit them. Normal human variations in body sizes and shapes exist within a relatively homogenous population, as evidenced by O'Connell's experiences in England and China. Genetic conditions such as dwarfism and a variety of neuromuscular conditions such as cerebral palsy or muscular dystrophy also add to the variability of human size and shape. Of course, traumatic injury such as the loss of a limb affects body size and shape. Human capabilities, both physical and cognitive, exhibit considerable variability. Products that will be used by or serve humans must reflect a knowledge of such inherent variability in the resultant designs.

Automobile manufacturers make extensive use of anthropometric data, both in their products and in their processes. Anthropometry is necessary for designing cars, vans, and trucks. For example, seating design and control-panel layouts are based on anthropometric data. However, the automotive industry also utilizes anthropometric data for workplace design by seeking to create safe work environments and processes that also yield high-quality products. For example, workstation design must include range-of-motion, reaching, and seating design considerations.

Automotive designers are critically concerned about maintaining and expanding their market shares. Ford Motor Company has initiated two novel design approaches, the results of which are now appearing in Ford vehicles. One is the Third Age Suit [6, 7], and the other a pregnancy simulator [8, 9].

The Third Age Suit is designed to limit the range-of-motion of head, neck, arms, and torso. It has gloves designed to reduce the sense of touch and restrict hand movement. Goggles are provided with lenses that simulate vision with cataracts. The intent is for young designers to experience some of the problems of the elderly and then design products that reflect their experiences by affording greater accessibility and usability for this market segment. This same intent warrants using the pregnancy simulator.

As its name implies, the pregnancy simulator simulates pregnancy for the wearer. The following is a quote from Paul Lienert, a reporter who tried out the pregnancy simulator:

> Last month, I strapped around my already ample midsection a Birthways Empathy Belly—better known as a "pregnancy suit." I couldn't take a deep breath. My lower back ached. My bladder screamed for relief. Worst of all, I couldn't climb out of the back seat of the 2004 Mercury Monterey minivan. ... I looked around desperately for a grab handle or some kind of leverage—cursing the short-sighted Ford Motor Co. engineers under my breath—before I finally figured out how to half-roll my engorged torso out of the Monterey's second-row bucket seat [8].

Because of experiences like this, the next-generation Monterey will be different. The Monterey's design team included 40 working mothers who had a strong input on final design. The new design includes adjustable seats, pedals, and mirrors; better placement of control panel and controls; grab bars for easier entrance and exit (front and back seats); and a power lift-gate, which management wanted to eliminate but the women considered essential [8].

At the April 2003 New York Auto Show, "Gene Stefanyshyn, who's in charge of midsize cars for General Motors Corp. showed off the 2004 Malibu Maxx hatchback and described how it was designed with pregnant women in mind. ... Historically, we've designed cars around the 50th percentile male figure" [8]. That is changing [10].

Chapter six: Ergonomically sound

The automobile industry is working very hard to make its products and services accessible and usable to as many people as possible. This push to universal and accessible design does not mean one product fits or serves everyone; rather, it means a spectrum of products and services that together span as many market segments as possible [11, 12].

The workplace is another environment where a person's body size and shape are important. OSHA requirements [3] directly focus on the work environment. The ergonomic concerns deal with excessive reaching, turning, rotating, picking, and carrying—physical activities that are safety concerns. Note that "excessive" or "harmful" are relative terms in that what is excessive in terms of reaching for a person who is 6 feet tall will be different than for a person who is 5 feet tall. An excessive load to pick up will be different for a 5-foot, 110-pound female than for a 6.5-foot, 200-pound male. Hence, a person's anthropometric features play a critical role in workplace design.

The capability to model a variety of body sizes and shapes for job process design and planning has become increasingly important in today's global economy where multinational corporations have manufacturing and assembly plants around the world. In that regard, it is noteworthy that Delmia, a leading supplier of process analysis and simulation tools, has a new product called ENVISION/ERGO™ [13]. This product is a human motion and task analysis tool to rapidly evaluate multiple worker/process scenarios. Analysis capabilities include range of motion, National Institute for Occupational Safety and Health (NIOSH) lifting guidelines, energy expenditure, upper-limb repetitive motion assessment, and Methods Time Measurement (MTM-UAS). Delmia's promotional material notes that by using ENVISION/ERGO™, engineers can create mannequins for different nationalities with different body types and shapes [13].

The *ADA Accessibility Guidelines (ADAAG)* for buildings and facilities, published by the Access Board [14], provides accessibility guidelines based heavily on ergonomic considerations. For example, Section 4.2 of the guidelines deals with space allowances and reach ranges [14]. Section 4.9 deals with stairs and has subsections specifying the width of stair treads (Section 4.9.2) and the placement of handrails (Section 4.9.4) [14].

Automated teller machines (ATMs), automated checkout systems, and information kiosks are examples of systems that should be accessible to people with a wide range of physical attributes: very short people, very tall people, people who are very thin to those who are obese, and people who use wheelchairs. Users should be able to reach the controls, see the displays, lift items into bags, present credit cards to automatic readers, and enter PIN numbers on a keypad or touchscreen. Hence, these service systems must be designed for sound ergonomic functioning for a wide spectrum of users.

The final rules for Section 508 which deal with electronic and information Technology (E&IT) also contains specific language that derives from anthropometric considerations [15]. Consider, for example, Section § 1194.25 dealing with self-contained, closed products, such as public kiosks. This section provides specific ergonomic requirements:

Where any operable control is 10 inches or less behind the reference plane, the height shall be 54 inches maximum and 15 inches minimum above the floor. Where any operable control is more than 10 inches and not more than 24 inches behind the reference plane, the height shall be 46 inches maximum and 15 inches minimum above the floor. Operable controls shall not be more than 24 inches behind the reference plane [15].

The key point of this example is the detailed specification regarding the positioning of operable controls.

Strategy 3: Design for ease of use

Given that a designed entity can operate without incurring any ergonomic risk factors, and that controls, doors, latches, and other manipulable elements are within reach, the forces and torques necessary to operate the entity must be within acceptable ranges. For example, levers are easier to use than doorknobs for people with muscle weakness. Another example is living space that allows for ergonomic ease of use.

Modern trends toward larger houses are creating additional demands for universally designed homes and living spaces. These trends are driven not only by broadly based consumer tastes, but also changing demographics [16, 17]. "Many boomers aren't planning to downscale after they retire, or at least not to the extent that their parents and grandparents did. Today's middle aged population is accustomed to larger homes. According to the National Association of Home Builders, the average size of new American homes has doubled in the last half century (from 983 square feet in 1950 to 2225 square feet in 1999)" [17]. Moreover, the boomers want to age in place and not have to move to meet changing needs and they do not want special accommodations to be obvious [17].

In general, this group wants "open floor plans, the master bedroom and bath on the first floor, so as to avoid the ergonomic challenge of stairs. Excessive reaching and bending could be problems due to arthritis; adjustable kitchen counters and storage space address these ergonomic challenges. Balance and walking problems are addressed when there are no differences in floor height between rooms and there is flush flooring for entrance ways (no molding or bumps). Such designs also provide greater walker or wheelchair accessibility, as do wider hallways and 30-inch (minimum) doorways. These designs make homes appealing and easier to live in without giving them an "old folks" look" [17].

Truck designers have worked hard to improve the ergonomic demands of truck use as the number of women driving trucks has dramatically increased. The force required to open tailgates, trunks, and doors has been significantly reduced. The position of door openers, key slots, and latches have been changed so that both men and women can easily reach them. Steps and hand grips have been added to make entering and exiting the vehicles easier for all users. All of these features exemplify design Strategy 3: design for ease of use.

The final rule for Section 255 of the Telecommunications Act contains design requirements for telecommunication products and services that derive from ergonomic considerations associated with ease of use [2]. For example, two regulations are § 1193.41 sections (e) and (f). These sections deal with input, output, and other product or system control functions. The controls must be:

> (e) Operable with limited manual dexterity. Provide at least one mode that does not require user fine motor control or simultaneous actions.
>
> (f) Operable with limited reach and strength. Provide at least one mode that is operable with user limited reach and strength [2].

These requirements target ease of use. They do not specify risk avoidance requirements nor detailed positioning requirements related to body size. They simply state that system control functions should be easy to use.

Conclusions

In conclusion, ergonomic considerations are critical for universal and accessible design approaches. Utilizing the three design strategies (design to avoid ergonomic risk factors, design for a wide range of body sizes and shapes, design for ease of use) covers the essentials of ergonomic engineering. Market forces dictate such consideration but federal laws also mandate accessibility for a wide range of facilities, products, and services.

As noted in the previous chapter, ergonomics, as defined herein, is only one facet of human factors. The next human factors facet to consider is perception.

References

1. Human Factors and Ergonomics Society, Human Factors and Ergonomics Society Home Page. Provides links to resources and definition of terms, http://hfes.org/, 2004.
2. Access Board, *Telecommunications Act Accessibility Guidelines: Final Rule*. Washington, D.C.: Federal Register, 36 CFR Part 1193, 1998.
3. Occupational Safety and Health Administration, "Ergonomics Program: Proposed Rule," *Federal Register*, 64, 65768–66078, 1999.
4. Merriam-Webster, *Webster's Ninth New Collegiate Dictionary*. Springfield, MA: Merriam-Webster Inc., 1985.
5. D. O'Connell, "Hong Kong is the Top Shop for Small People," Guardian Newspapers Limited, 2002.
6. Loughborough University, "The Third Age Simulation Suit," http://www.lboro.ac.uk/taurus/simulation.htm, 2001.

7. A. Lienert, "Mobility Challenges: Ford engineers age quickly in bid to woo older buyers. Outfit that simulates aging helps adapt designs to people with arthritis, bad backs and failing eyesight," in *The Detroit News*; http://detnews.com/2005/specialreport/0501/09/B06-54376.htm. Detroit, January 9, 2005.
8. A. Lienert, "Designers have moms on the mind: new vehicles include more friendly features," in *The Detroit News*. Detroit, May 11, 2003, pp. 1B, cont. 4B.
9. P. Lienert, "Driving pregnant is far from bliss," in *The Detroit News*. Detroit, May 11, 2003, pp. 1B.
10. Ergonomics Society Press Release, "Ergonomics in Car Design Benefits Older Drivers," http://www.ergonomics.org.uk/press/press2.htm, 1999.
11. J. P. Womack, D. T. Jones, and D. Roos, *The Machine that Changed the World*. New York: Harper-Collins Publishers, 1990.
12. J. P. Womack and D. T. Jones, *Lean Thinking*. New York: Simon & Schuster, Inc., 1996.
13. Delmia, Delmia Home Page. Provides links to products and product descriptions, http://www.delmia.com/, 2004.
14. Access Board, "ADA Accessibility Guidelines for Buildings and Facilities (ADAAG)," http://www.access-board.gov/adaag/html/adaag.htm, 2002.
15. Access Board, *Electronic and Information Technology Accessibility Standards*. Washington, D.C.: Federal Register, 36 CFR Part 1194, 2000.
16. Design Basics Inc., "The Best Bathroom Ideas," http://www.designbasics.com/homebuyers/designtrends-19.asp?FormUse=homebuyers, 2004.
17. L. Reimer, "New Homes for Aging Boomers," http://www.designbasics.com/homebuyers/designtrends-10.asp?FormUse=homebuyers, 2003.

chapter seven

Perceptible

Chapter goals:

- State the perception universal design principle and associated design strategies.
- Show how human physiological, psychological, and psychometric capabilities shape universal design strategies associated with human perception.
- Provide examples of applying the design strategies.

Perceptible

Principle

Designed entities must effectively communicate necessary information to the user, regardless of ambient conditions or the user's sensory abilities [1].

Discussion

A traveler using her cell phone in the airport has trouble hearing her caller because of the loud ambient background noise in the terminal. Much of accessible design deals with perceptual issues, overcoming ambient noise, or providing adequate tactile feedback. Emergency warning systems are prime examples of systems that must be perceptible by as many people as possible. To overcome unknown and unpredictable ambient conditions, the sounds are loud, employing an attention-getting pitch alternating in intensity or pitch. The lights are bright, vivid, and most often flashing.

Emergency warning signals are "in your face" or "ears," as the case may be of perception. The vast majority of perceptual issues are more subtle and so common that the design solutions seem obvious and of little concern to designers. Such familiarity can lead to a "cookbook" type of design solution, repeating previous solutions and not considering the possibility of new innovative design approaches. The temptation is to treat perceptual issues without a real understanding of the underlying physiological processes that drive perception.

This chapter explores some of the physiological processes that underlie perception and how these processes lead to specific universal and accessible design strategies. The intent is that knowledge gained from such understanding will lead to more innovative design solutions to perceptual issues.

Design strategies

The design strategies start with the most obvious perceptual issue: can a person sense the signal? Is the person deaf or hearing impaired, or in a very noisy environment? Is the person blind or temporarily visually impaired because he is in a dark room, or blinded by the sun or bright flash of light? If a person cannot sense a signal or message via a given sensory modality, then the widely used design strategy is to provide multiple sensory input channels. If the environment is noisy, dark, or in some way hinders one's ability to perceive a signal or message, then whenever possible the user should be able to turn up the volume, increase the visual contrast, or make some adjustment so as to increase perception. Strategies 4 and 5 address these perceptual issues.

Strategy 4: Provide multisensory options for communications between a person and the process or product

For example, provide both visual and auditory prompts, messages, or signals. For signage, maps, icons, and other information or communication devices, do not use only static color-coding schemes, but also include aids that do not rely solely on color (for example, words, icons, arrows, or symbols). Such multimedia approaches address issues of sensory impairments or loss by providing multiple means of expressing the same message.

Section 255 of the Telecommunications Act of 1996 provides some examples of this design strategy. One example is from § 1193.43 Output, display, and control functions:

> All information necessary to operate and use the product, including but not limited to, text, static or dynamic images, icons, labels, sounds, or incidental operating cues, shall comply with each of the following, assessed independently.
>
> (a) Availability of visual information. Provide visual information through at least one mode in auditory form.
> (b) Availability of visual information for low vision users. Provide visual information through at least one mode to users with visual acuity between 20/70 and 20/200 without relying on audio. ...

(d) Availability of auditory information. Provide auditory information through at least one mode in visual form and, where appropriate, in tactile form [2].

§ 1193.43 ensures that the information necessary to operate output, display, and control functions is provided in more than one sensory mode, in particular in visual, auditory, and Braille formats.

Section 508 of the Rehabilitation Act also has a variety of accessibility requirements for electronic and information technology based on this design strategy. The following are examples of such requirements. § 1194.25 deals with self-contained, closed products. One subsection states that "(g) Color coding shall not be used as the only means of conveying information, indicating an action, prompting a response, or distinguishing a visual element" [3]. Another example comes from § 1194.31, which deals with functional performance criteria. This section has the following language; the system shall have "[a]t least one mode of operation and information retrieval that does not require user vision shall be provided," and also "[a]t least one mode of operation and information retrieval that does not require user hearing shall be provided" [3]. All these cases require multisensory options for communications between a person and the process or product.

Another example of this strategy can be found in Section 9 Accessible Transient Lodging of the *ADA Accessibility Guidelines (ADAAG)* for buildings and facilities, published by the Access Board [4]. Section 9.1.3 specifies the number of accessible rooms for persons with hearing impairments. These rooms require auxiliary visual alarms and notification devices for the telephone (Section 9.3), that is, providing multiple sensory alarms so as to ensure the alarm is sensed.

In the work environment, the process or workplace must effectively communicate necessary information to the worker, regardless of ambient conditions or the worker's sensory abilities [1]. The design should utilize redundant modes of information presentation (for example, verbal, iconic, pictorial, tactile). The design should maximize legibility by providing adequate contrast between the information and its surroundings.

Strategy 5: Design signals so as to maximize the signal-to-noise ratio

Each physical signal has an energy content that is measurable. Sound pressure is measured in decibels, light in lumens, and touch in force. For the signal to be perceptible, the signal's energy level must be greater than the energy level of any ambient or background noise. Noise in this context means any energy not derived from the signal. The signal-to-noise ratio is the ratio of the signal's energy content to the noise's energy content. The higher the ratio, the more signal energy is present relative to noise energy.

Providing lights to illuminate an area is a classical example of this strategy. Public streetlights are turned on when the sun goes down. Interior lighting is provided for buildings. Flashlights are used for mobile illumination. Volume controls on public address systems, car radios, or cell phones are additional examples. However, these very common examples obscure the complexity of the physiological processes underlying the design solutions. Considering the underlying physiology of signal detection can further refine strategy.

The following strategies derive from the signal detection and information processing characteristics and capabilities of humans. A collection of psychometric laws has been formulated to describe signal detection and perception. Weber's law, Fechner's law, and Stevens' power law are three examples of such laws. Refined design strategies are derived from these psychometric laws.

Signal detection

Sensory discrimination is based on signal detection. If the signal is not detected, then there is no discrimination. Different physical stimuli are perceived and processed in different ways by the nervous system. Understanding how humans process signals and sensory stimuli is important in the design of accessible and usable products and systems.

Weber's law, Fechner's law, and Stevens' power law all deal with human signal detection. A physical stimulus in the physical world possesses a certain amount of energy; it has a physical intensity. That signal reaches a physiological detector, for example, a pressure receptor in the skin, a sensing element on the retina of the eye, the semicircular canal of the ear, an olfactory receptor, or a taste receptor. At the receptor, a signal transformation occurs, one form of energy (from the physical world) is transformed into electrochemical energy at the receptor, and then nerve impulses make their way to the information processing centers of the brain. When the internal nervous signals are processed, a signal or stimuli is perceived.

Weber's law

Weber and Fechner studied the relationships between the physical intensity of a stimulus or signal and the perceived psychological intensity of the physical stimulus. They did this by asking human subjects to compare the relative physical intensities of stimuli and then report on the psychological intensity.

For example, if a person is carrying a 40-pound television set and someone places a small screwdriver on it, the person would not notice the added weight. However, if someone places a 10-pound box of books on the TV, the additional weight would be noticed. Experiments were conducted wherein a subject was asked to lift a 5-kilogram weight. Additional weight is incrementally added until the subject reports a noticeable difference. This might be 0.1 kilograms for an average-sized man. Weber and Fechner noticed that if the weight started at 10 kilograms, the "just noticeable difference" (JND) was 0.2 kilograms.

Fechner coined the term "Weber's law" to model the generalized principle: The ratio of the "just noticeable difference" to the original intensity is a constant. Mathematically, $\Delta I/I = K$.

The formula $\Delta I/I = K$ says that the larger the signal, that is, the larger the magnitude of the stimulus (I), the larger the change in magnitude (ΔI) must be to be detected. For example, if someone whispers quietly in a library, that person will be heard; but the same individual must yell to be heard over the roar of a crowd at a football game.

Fechner's law

Fechner expanded on Weber's work by postulating a model of how the perceived psychological intensity is related to the physical intensity. This is a tricky proposition given that subjective intensity cannot be directly measured. Fechner's law emphasizes the compressed representation of the physical stimuli in the psychological domain. This compression implies that when varying signals or stimuli, it may be necessary to make relatively large changes in the physical stimuli to render it psychologically perceptible.

Examples of Fechner's law include using volume controls on radios, televisions, telephones, and devices with auditory outputs. Other examples are the screen display features that allow computer users to select text size, background color, and contrast options for their displays.

While Fechner's law holds for a wide range of stimuli, it does not cover all situations. For example, Fechner's law does not hold for certain line length estimations wherein the perceived length changes precisely with actual length, and it does not hold for perceived or experienced pain that grows faster than the pain stimulus. This is where Stevens' power law comes into play.

Stevens' power law

Stevens' power law generalizes Fechner's law and provides a model that includes not only the compressed representation of sensory processing, but also the 1:1 cases exemplified by the line length estimates, and the amplification representation exemplified by pain. Experiments based on Stevens' power law demonstrate that different stimuli are processed differently in terms of sensory processing. Table 7.1 provides examples of sensory stimuli from each of the three representation categories: compressed, 1:1, and amplified.

When designing prompting signals or messages, feedback, or warning signals, Stevens' law suggests using stimuli that produce perceived sensations more intense than the physical stimuli, that is, an amplification representation. Electric shock and other pain signals fall into that category; but while useful from a survival point-of-view, they are not suitable as a public warning or display stimuli.

The bottom row in Table 7.1 shows several sensory stimuli that exhibit amplified representation for perception. Tactile roughness is effective when

Table 7.1 Categorization of Stimuli According to Stevens' Formula

Representation of sensory processing	Sensory stimuli
Compressed representation (Fechner's law)	Loudness, vibration on finger, brightness in the dark, taste (saccharine), smell (heptane)
1:1 representation	Visual length (line), brightness (point source briefly flashed), cold (metal contact on arm)
Amplified representation	Electric shock, redness, tactile roughness, brightness (brief flash), taste (sucrose)

Source: From Class Notes, Stevens' Power Law, [11].

used as rumble strips on the highway to warn drivers to slow down, or that they are on the side of the road. Tactile roughness or other tactile cues can be used on control surfaces such as knobs, touch panels, or switches to help users identify such entities. Tactile roughness is often used on approaches to stairs and sidewalks to warn of stairs or a curb.

It is not surprising that flashing warning lights are often red lights in that these stimuli, redness and brightness (brief flash), also possess an amplification representation. Police, emergency vehicles, and ambulances use flashing, bright lights and very loud noise to get people's attention. The noise needs to be loud because it has a compressed representation.

Designers can use every human sense as a method of communication. Sound, vision, and tactile stimuli have already been discussed. Smell is used as a warning for natural gas in that an additive provides the unique smell we call "gas." Food quality is a multisensory phenomenon. Taste is certainly a critical component. Chefs, cooks, and other food designers are critically concerned with taste and design specific dishes to have specific tastes. The sense of smell and taste are central in assessing whether food has spoiled.

Weber's law, Fechner's law, and Stevens' power law lead to two strategies that target the need for large signal-to-noise ratios under different environmental conditions.

Strategy 5.1: Provide the ability for a person to increase or decrease the signal strength so as to increase the signal-to-noise ratio

One can control the sound volume on a car's radio to compensate for traffic and road noise. One can adjust the brightness on a TV screen to compensate for room illumination. One can adjust the vibration strength on a pager or cell phone. In all of these examples, the user could increase the desired signal strength so as to overcome ambient or background conditions.

Paragraph § 1194.25 (f) of Section 508 of the Rehabilitation Act Amendments of 1998 states that "When products deliver voice output in a public area, incremental volume control shall be provided with output amplification up to a level of at least 65 dB. Where the ambient noise level of the environment is above 45 dB, a volume gain of at least 20 dB above the ambient level shall be user selectable. A function shall be provided to automatically reset the volume to the default level after every use." This legal requirement derives from Weber's law.

The following example is from Section 508 of the Rehabilitation Act. Subsection § 1194.21 covering "Software applications and operating systems" specifies that "(j) When a product permits a user to adjust color and contrast settings, a variety of color selections capable of producing a range of contrast levels shall be provided" [3]. This language recognizes the compressed representation of sensory processing as stated in Fechner's law and prescribes design guidelines to address this phenomenon.

Consumer products use adjustable volume controls, contrast, and color controls on computer monitors or televisions, and text and font selection mechanisms on computer displays. All these adjustable control features allow the user to adjust the sensory input to ensure adequate perception. Recall that Section 508's Subpart C § 1194.31 Functional performance criteria (from Chapter 4) covers this topic with respect to the accessibility requirements for federal agency purchases. Public kiosks should have adjustable volume controls to help users overcome loud ambient noise. The design of public kiosks or information systems should utilize redundant modes of information presentation (for example, verbal, iconic, pictorial). If the kiosks are for use in federal facilities, they must provide such adjustability capabilities [3].

> *Strategy 5.2: Provide sufficient contrast between signals and background ambient conditions so that most people will be able to perceive the signal and the message conveyed by the signal*

If the user cannot control the signal or ambient conditions as with an exit sign, public announcement, or warning signal, then the designer must select signals that provide an adequate signal-to-noise ratio under virtually all ambient conditions. Using bright flashing lights, loud piercing sirens, and high contrast between letters and background on an exit sign all exemplify design Strategy 5.2.

Another approach is to have the system sense the ambient conditions and automatically adjust the signal-to-noise ratio. For example, the digital display on the Bose® Wave Radio/CD dims when the room lights are turned off and brightens when ambient lighting increases.

The growing use of sound field systems by schools is another example of universal design to address perceptual concerns. The teacher wears a cordless microphone and a portable amplifier/transmitter that transmits a signal to a receiver that then distributes the signal to speakers placed around the classroom. The sound field technology started in special education classes but research shows that at any given time, a high percentage of general education classroom students have hearing problems due to colds, flu, allergies, fatigue, and other temporary conditions [5]. Research also shows that students in sound field classes perform better academically than those not in such a class [5]. Hence, there is a growing trend to install sound field technology in all classrooms.

Consider again emergency signals such as the backup signal on trucks, high-lows, and other mobile work equipment. They are designed to be loud and perceptible. Working robots are typically surrounded by a sensor array. If anyone enters the work envelope of the robot, light and sound warnings are activated. Exit and warning signs are high contrast with intense colors for the wording. Walk and work areas in manufacturing plants are designated by color-coded floor markings.

Signal detection as expressed by Weber's law, Fechner's law, and Stevens' power law provide the rational for two common design strategies. While adherence to these design strategies may ensure signal detection by a person, the signal is typically intended to convey a message, i.e., carry information to the person. In terms of using products or services or performing tasks in a work environment, people are required to make judgments and perform actions. The next subsection considers judgments that are based on perceptual processes.

Judgments

People have to make judgments based on sensory perceptions all the time. When driving, how much space should I leave between my car and the car in front of me? Can I safely merge into the expressway traffic flow from the entrance ramp? How long is that shelf? How does the color on the paint chip sample match the color of the paint on the wall? Are the bass, treble, and balance of the room's audio system okay? Was that audio signal (Morse code) a "dash" or a "dot"?

Work environments provide additional judgment challenges. Any product inspection process involves a judgment: Is this cord the correct length? Is this the correct weight? A doctor examining an x-ray image makes judgments based on the observed density, lines, and patterns. An engineer makes judgments regarding temperatures from a pseudo color rendering of an infrared remote sensing device.

Research on human judgment has discovered some fundamental principles related to the information processing capabilities of human physiological systems. For example, people are not good at making absolute judgments. People are much better at making relative judgments. The following experiment by Pollack, where listeners were asked to assign numbers to tones, exemplifies experiments of absolute judgment [6, 7]. In this experiment, only the tone's frequency was changed and the frequency varied from 100 to 8000 Hertz in equal logarithmic steps. The lower frequencies were to be assigned lower numerical values. For the experiment, a tone was presented and the listener responded by giving a number. The subject was then informed of the correct identification. As the amount of input information increased, by increasing the number of different frequencies to be judged from 2 to 14, the data showed that a listener cannot identify more than six different pitches accurately. If listeners are asked to form judgments for more

Table 7.2 A Comparison of the Discrimination Capacities for Different Senses and Tasks

Sense	Task	Alternatives that can be discriminated
Auditory	Absolute judgments of tone loudness	≈5
Taste	Absolute judgment of saltiness	≈4
Visual	Absolute judgment of the position of a pointer in a linear interval	≈10
Visual*	Two-dimensional positioning of a dot	≈24
Auditory*	Loudness and pitch of pure tones	≈9
Visual*	Colors of equal luminance	≈12

* Denotes multiple stimuli.

Source: From G. Miller, The Magical Number Seven, Plus or Minus Two: Some Limits on Our Capacity for Processing Information, *The Psychological Review*, 63, 81–97, 1956.

than six equally likely pitches, they will make errors by grouping the presented pitches into about six categories. Of course, such conclusions depend on the person; musically sophisticated persons with absolute pitch have been know to accurately identify any one of 50 to 60 different pitches [8].

Table 7.2 provides a comparison of the discrimination capabilities of different senses and tasks. It is clear that there is an interaction between the task and the sensory system's discrimination.

The experiments and data on absolute judgment indicate that increasing the number of alternatives, for example, more levels of brightness, loudness, pitch, etc., eventually leads to problems due to saturation of the information processing channel. The experimental evidence suggests that people start making mistakes of absolute judgment when the number of equally likely alternatives for a specific type of signal exceeds a threshold value associated with the signal and sensory modality. Experimental data also suggests that multiple stimuli, multiple signals over different sensory processing channels, increase the total signal processing capacity. However, this increased capacity comes at a cost, a decrease in the judgment accuracy for any particular signal.

Clearly, people start to confuse signal judgments when the number of levels, pitches, or, in general, the absolute judgments become excessive. Experimental results suggest that excessive means more than five or six. Therefore, the designer needs to keep the signal detection requirements simple. This leads to another design strategy.

Strategy 6: Keep the signaling structure and content as simple as possible

Most people cannot distinguish more than about six different signal levels before they get the signals mixed up. Morse code is a very successful messaging protocol. One of the reasons it works so well is that it uses simple dots and dashes, not a complex sequence of very short, to short, to medium, to long, to very long signal segments. People would be confused by such a complicated protocol.

The AT&T™ 9350 Digital Answering System Speakerphone telephone answering machine uses ten different audible signals, ringing, beeps, and tones to communicate different information. For example, long ringing signals an incoming call, four short beeps signal a low battery, a three-part tone signifies a page, a series of ascending tones indicates that an action has been successful. People will forget what the rarely used signals mean; there are just too many. Also, some people might have trouble distinguishing among the signals (for example, three or four beeps). This could be a perception problem or a memory problem.

As simple as possible includes the use of techniques that take advantage of people's strengths, not weaknesses. Human work experiences supported by the absolute judgment experimental data suggest that an absolute judgment approach is highly error-prone and unreliable as a quality control procedure [9]. The process engineer can design a much more effective quality control strategy. The most straightforward strategy would be to have the inspector compare the new product to a standard or reference product, a relative judgment [9]. If the job is to determine if an item is of proper length, then a proper length reference for comparison should be provided. This technique turns the absolute judgment task into a relative judgment.

Reaching to touch or push a button, picking up objects, or placing an object in a slot or on a table are all examples of judgments that involves the motor system and perception. We must move our hand, arm, leg, whatever to a particular point in space. Such tasks are complex because of the interactions between the perceptual systems and the body's motor system. The next section considers Fitts' law. It is one of many laws related to psycho/motor phenomenon in humans. An understanding of these laws provides the basis for universal design strategies, which seek to avoid design solutions that are fundamentally and universally difficult for humans in general and utilize design solutions that take advantage of human strengths.

Movement: Reaching, picking, and placing

Reaching to pick up an object, placing a credit card into a card reader, or dialing a number on a cell phone are all activities that involve reaching out with a controlled, aimed movement. Fitts' law can model such target acquisition activities. Fitts' law can be summarized as: the bigger the target, the easier it is to hit. Fitts' law leads to another design strategy.

Strategy 7: Design entities so that a user can accurately acquire a target

Examples of Fitts' law can be found in processes such as placing groceries into a shopping bag, or inserting an ATM card into the receiving slot. Other examples include the control keys on TV and VCR remote controls. The size and placement of the pushbuttons make operation more or less difficult.

Chapter seven: Perceptible

Figure 7.1 Two remote control units. The one on the right was designed to be more accessible. The device is larger with larger control buttons that are back-lit when a button is depressed.

Figure 7.1 shows two remote control units. The larger one on the right was designed using universal design principles. The device itself is larger, making it easier to hold the unit and be able to see and operate the buttons. The control buttons are large and when a button is depressed, the buttons are backlit, thus providing additional contrast for button identification. The unit on the right in Figure 7.1 is much easier for elderly persons, or persons with big hands or fine motor difficulties such as arthritis of the hands or cerebral palsy.

The influence of Fitts' law also appears in ADA facilities regulations. One example is from Section 4.10.12 of the ADAAG, which covers controls in elevators and states that the control buttons on elevator control panels "shall be at least ¾ inches in their smallest dimension" [4]. This is meant to ensure that people can easily push the elevator controls.

The assembly of components on an automobile has been explored extensively with the slogan "Tight Targets Take Time." Productivity and quality are of great concern to the automotive companies. An assembly that requires four screws with associated washers and lock nuts is much slower than an assembly that uses screw-less fasteners. The screws must be inserted. The washers and lock nuts must be attached. These operations exemplify Fitts' law. The process of using a power screwdriver also requires precise placement of the screwdriver tip into the notch on the head of the screw. These are all relatively slow operations because they are "tight tasks." The placement of screw-less fasteners does not require all the target acquisitions associated with the use of screws.

Design for assembly (DFA) is a design methodology that addresses such assembly concerns [10]. DFA incorporates an assembly assessment process and methods for identifying where the assembly has steps that are too time-consuming or error-prone [10]. Once flagged, the designer can rethink the original design and remove the problematic features.

Conclusions

The fundamental premise of this chapter is that an understanding of human physiology, psychology, and psychomotor capabilities leads to universal design strategies. They are universal design strategies in that design solutions can avoid those elements that are fundamentally difficult for humans because of our human endowments and include elements that people are good at because of our common physiology.

Physiological impairments brought on by any one of a number of reasons, from traumatic injury accidents, to birth defects, genetic disorders, and the aging process, affect one's perceptual abilities. Universal and accessible design strategies also address the negative impact of such impairments by seeking design solutions that eliminate or reduce the disabling effect of the designed environment on the person.

Two facets of human factors have now been considered, the final facet deals with cognitive design issues.

References

1. B. R. Connell, M. Jones, R. Mace, J. Mueller, A. Mullick, E. Ostroff, J. Sanford, E. Steinfeld, M. Story, E., and G. Vanderheiden, *Principles of Universal Design*. North Carolina State University: The Center for Universal Design, 1997.
2. Access Board, *Telecommunications Act Accessibility Guidelines: Final Rule*. Washington, D.C.: Federal Register, 36 CFR Part 1193, 1998.
3. Access Board, *Electronic and Information Technology Accessibility Standards*. Washington, D.C.: Federal Register, 36 CFR Part 1194, 2000.
4. Access Board, "ADA Accessibility Guidelines for Buildings and Facilities (ADAAG)," http://www.access-board.gov/adaag/html/adaag.htm, 2002.
5. C. Crandell, J. Smaldino, and C. Flexor, "Sound Field Systems Improve Learning," *The Hearing Rev.*, 6, 40–42, 1999.
6. I. Pollack, "The Information of Elementary Auditory Displays," *J. Acoust. Soc. Am.*, 24, 745–749, 1953.
7. I. Pollack, "The Information of Elementary Auditory Displays. II.," *J. Acoust. Soc. Am.*, 25, 765–769, 1953.
8. G. Miller, "The Magical Number Seven, Plus or Minus Two: Some Limits on Our Capacity for Processing Information," *Psycholog. Rev.*, 63, 81–97, 1956.
9. N. K. Shimbun, *Poka-Yoke: Improving Product Quality by Preventing Defects*, Cambridge, MA: Productivity Press, 1988.
10. G. Boothroyd and P. Dewhurst, *Product Design for Assembly*, 3rd ed. Wakefield, RI: Boothroyd Dewhurst, Inc., 1991.
11. Class Notes, "Stevens' Power Law," http://www.cis.rit.edu/people/faculty/montag/vandplite/pages/chap_6/ch6p10.html, 2003.

chapter eight

Cognitively sound

Chapter goals:

- State the cognitively sound universal design principle and associated strategies.
- Present psychological and physiological bases for the specified strategies.
- Show how to assess complexity and create simpler design.
- Provide examples applying the design strategies.

Cognitively sound

Principle

The cognitive demands of designed entities must be within acceptable limits for a wide range of users.

Discussion

The final rules of Section 255 of the Telecommunications Act of 1996 has a section devoted to cognitive accessibility. The language is interesting in so far as it is the least detailed in terms of providing guidelines.

> **§ 1193.41 Input, control, and mechanical functions**
>
> (i) Operable with limited cognitive skills. Provide at least one mode that minimizes the cognitive, memory, language, and learning skills required of the user [1].

Cognitive demands include, but are not limited to, memory, language, and learning requirements, as well as task complexity. Designers have long addressed the problems of cognitive demands by building knowledge into the environment and striving for designs that reduce complexity.

The cognitive demands of our everyday lives are significantly reduced by knowledge in the environment. Traffic lights and road markings are common examples of knowledge built into the environment. Another example is the variety of navigation aids used in large public buildings such as mega-stores. In a recent newspaper article discussing how Meijer plans to compete with Wal-Mart, a photograph shows the spacious aisles with a variety of ceiling-hung and floor-mounted signs. The caption reads: "Among the features of Meijer's newest store in Rockford are signs directing shoppers to the appropriate aisles" [2]. The thrust of Meijer's is convenience. Meijer's store designers want people to easily and quickly get in and find what they want. Color-coding schemes and path markers are commonly used. In public places around the world, international icons and symbols are being used to inform the public as to the location of restrooms, restaurants, lodging, hospitals, stairs, elevators, and escalators. These internationally recognized icons allow people of differing nationalities as well as those who cannot read or are cognitively impaired to negotiate complex environments.

The cognitive demands required for product use can be reduced if designers of consumer products build knowledge into their products. Using international icons for electronic recording and playback equipment exemplifies building knowledge into the product. Using mappings (the layout of the burner and oven controls on a kitchen range) provides another example of building knowledge into the environment. Finally, using on-screen menus for VCR programming greatly simplifies programming these devices and significantly reduces the cognitive load on the user.

Automotive designers have done a good job of building knowledge into the car's internal environment. Typically over 120 different controls exist in a car, yet most people can get into a new car and without much trouble figure out where the controls are and how they work [3]. Automotive dashboard controls extensively use international icons to identify control functions for doors, windows, lights, turning, defrosters, heating/cooling, etc. These icons have broad cultural semantic understanding. The controls make extensive use of natural mappings; for example, push the window control down to lower the window, up to raise the window. People would have much more trouble trying to figure out the controls of a military jet fighter. In the jet, people cannot rely on the cultural and semantic conventions employed by the automotive designers and understood by the driving public to help users understand how everything works.

The cognitive demands associated with the use of a telephone system for consumer support and related service activities presents a unique set of design challenges. The cognitive demands arise with respect to memory, response time, alternative choices (TTY or real person), boredom, and frustration. Users must receive detailed auditory instructions as to how to proceed. If these instructions are too long, users may forget some by the time all options are given. If users have trouble hearing or understanding the instructions, they may not act soon enough to complete a transaction, or may simply guess at what they are supposed to do and risk making an error.

Chapter eight: Cognitively sound 97

Finally, users may wait so long for a response or person that they get frustrated and start pushing keys just to get attention or sabotage the process.

Looking at universal design in the workplace, the cognitive demands of work activities must be within acceptable limits for a wide range of workers. The task structure should be appropriate to the task—neither too simple (leading to boredom) nor too complex (leading to error and frustration). The material must support different learning styles and human intelligence [4]. Reading, performing mathematical operations, or memorizing long sequences of actions or codes places constraints on who can perform a job. An analysis of the cognitive demands of jobs is more difficult than an analysis of physical activity. Workers who cannot read, who are color blind, or who cannot remember code or action sequences may develop complex strategies to hide this from an employer. Typically, cognitive issues are noticed indirectly by increased errors, a decrease in productivity, and other symptomatic indicators.

Design strategies

Building knowledge into the environment, the product, or the process is essential to the design of cognitively sound entities [5, 6]. Knowledge in the environment comes from the affordances of an entity, and affordances arise from utilizing mappings and constraints. Error management is an essential part of cognitive soundness. Error management relies heavily on forcing functions, feedback, and warnings. In addition to these, designers must make good use of representations and language.

Strategy 8: Build knowledge into the designed entity or environment

This is the most fundamental design strategy with respect to the cognitive demands of using an entity or participating in a task. People rely on knowledge in the environment all the time—for example, traffic signals, signage, icons on the control knobs of radios, recorders, and other electronic devices.

Four design elements are generally associated with good cognitive design: (1) affordance, (2) mapping, (3) constraints, and (4) feedback [3, 5]. Each of these design elements can support human capabilities and hence support universal and accessible design strategies. Each of these elements can be used to build knowledge into the environment and thereby enhance usability. These elements should be considered basic building blocks for a usable design.

Affordance

Strategy 8.1: Use affordances to help users form clear conceptual models of the entity's operations

"Affordance refers to the actual and perceived attributes of a product or process that suggest its uses" [3]. Figure 8.1 shows a stool. Affordances

Figure 8.1 Stool.

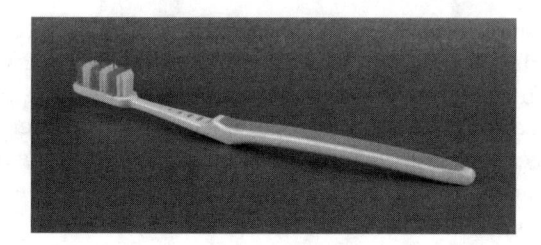

Figure 8.2 Toothbrush.

include a flat top, height of about 2 feet, and sturdy legs, which suggests that one can sit on this device, or place items on the flat surface, or stand on it to reach things. Figure 8.2 shows a toothbrush. The toothbrush represents a different kind of device. People know it is a toothbrush because of semantic and cultural knowledge. Someone from 1000 years ago might not realize that it is to be used to brush one's teeth. There are bristles on one end and a relatively long handle. The actual attributes suggest using it to brush things. The perceived attributes are culturally and semantically based and inform the contemporary person that this is a toothbrush. Figure 8.3 shows scissors. This device has affordances that suggest picking with the sharp metal points, cutting with the sharp metal edges, or using it as a weapon. Users' intuitive understanding of physics (simple machines) provides a conceptual model of how the scissors work, and this conceptualization is part of the device's affordance. Likewise for the stool in Figure 8.1, users' intuitive understanding of physics allows them to form a conceptual model that informs them that the device is sturdy enough to sit or stand upon without it breaking or tipping.

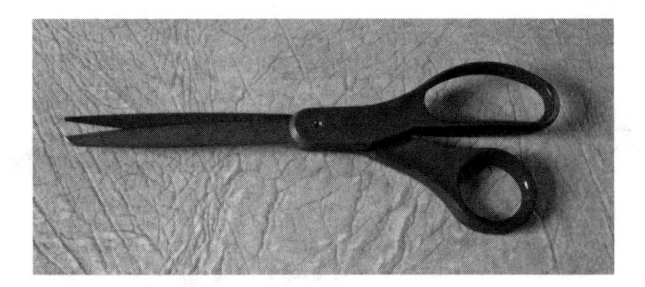

Figure 8.3 Scissors.

Affordance is determined by an object's design. Most of the time, designers want the affordance of an object to truly represent the object. Sometimes, designers want to fool the user, in the instance of plastic made to look like wood or metal. Other times, designers hide affordances, as in modern glass doors where the hinges are hidden and there are no obvious clues as to which side to push or possibly pull.

Affordance can provide a user with ergonomic operational information so as to help that person figure out how to use a device. Donald Norman has a classic discussion of doors in his book entitled *The Design of Everyday Things* [3]. Often it is not clear whether users push or pull a door. Door knobs suggest turning, whereas large faceplates invite pushing. It is a questionable design when a large, flat plate on a door has the word "PULL" on it.

Recently in a student design laboratory, a student pulled a control knob off the front panel of an oscilloscope. He explained that he thought the knob moved in and out a notch to provide different control functions. There were no markings to indicate such operation, but that is how the unit in his previous laboratory operated. This incident shows what can happen if conceptual models of ergonomic activity do not match reality—accidents or errors can and will occur.

Mappings and representation

Strategy 8.2: Use mappings to help users form clear conceptual models of the entity's operations and simplify operations

Mappings represent another important source of knowledge in the world. Norman defines a mapping as "the relationship between two things" [3]. Using natural mappings, a designer takes "advantage of physical analogies and cultural standards" to provide an immediate understanding. For example, turning a car's steering wheel to the right to turn the car right or moving a robot controller up to move the robot arm up illustrates natural mappings. Widely accepted cultural natural mappings exist, such as a rising level means more. Likewise, a louder sound might represent a greater amount or increase. Representations such as loudness, weight, line length, density, and brightness are additive dimensions and therefore naturally represent increases or decreases. Frequency or pitch, location, taste, and color are not additive

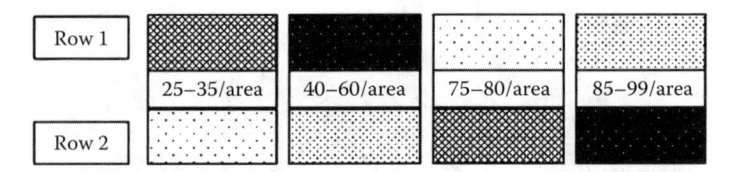

Figure 8.4 Example representations. The numbers represent some elements per unit area. Row 1's representation is less intuitive than Row 2's representation because in Row 2 the visual density increases as the numerical density increases.

dimensions; these entities are substitutive dimensions, that is, one representation is simply substituted for another with no particular rationale or logic for the substitution [5].

Mappings or representations should facilitate, not hinder, the user's knowledge-building process. For example, Figure 8.4 shows two ways of representing numerical density data. In Row 1 there is no correspondence between the numerical density data and the visual shading density. This is a substitutive representation in that a given shading represents certain numbers. In Row 2 the shading deepens as the numbers increase; thus, there is a relationship between the shading and the data. While Row 2 still substitutes a shading pattern for a range of numbers, there is a natural relationship between the density of the shading and the magnitude of the numbers. Row 2's mapping is more intuitive than Row 1's mapping because the visual shading becomes denser as the numerical density data it represents increases. The representation in Row 1 forces one to think about and compare the visual shading to the numerical data values; there is no natural mapping between the two representations.

Color is not an additive representation; it is a substitutive representation. A number of standard or widely accepted color palates exist. Figure 8.5 shows an infrared image of the human ear. The substitutive color scheme is called the rainbow medical color palette. On this palette the colors range

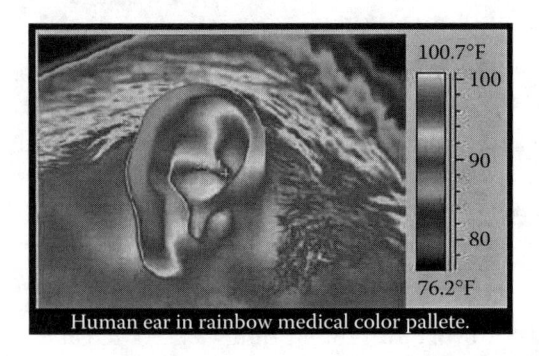

Figure 8.5 Infrared image of the human ear. (Courtesy of SPI Corp., www.imaging1.com.)

Chapter eight: Cognitively sound 101

from white (100.7°F) to deep purple, almost black (76.2°F). The color gradation relies on semantic and cultural knowledge to provide meaning to the color-coding scheme, with reds being hot and blues cold.

People have a limited information processing channel capacity and therefore designers must be careful to not design signaling that lends itself to misinterpretation, errors, or accidents [7, 8]. If the signals are too close in color, sound, pitch, or whatever the sensory signal, resolution and discrimination are lost. In Figure 8.5, most people cannot distinguish between the various shades of red around the ear and hence cannot be specific about the actual temperature. The region is in the mid to upper 90°F but not much more can be said. For general medical use, this level of resolution is acceptable; but for many manufacturing processes, this level of resolution would be unacceptable. If a person is color blind, then color-coding is of limited or no value. Likewise, if a person is deaf, then an auditory cue is meaningless.

Another example of a mapping is the layout of burner controls on a kitchen range. Figure 8.6 shows a range and its associated control panel. Sometimes the manufacturer provides a simple key indicating which burner the knob controls. Other times one simply assumes a correspondence between controller location and the burner it controls. In this example, the location of the control informs the user—left-side control for the left burners,

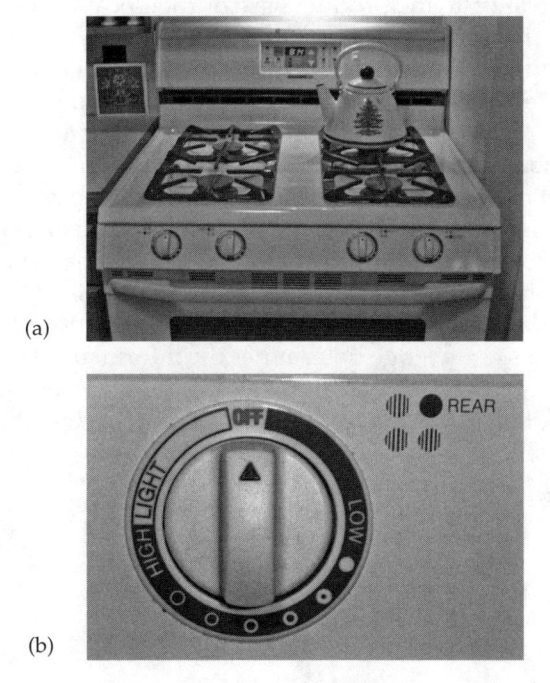

(a)

(b)

Figure 8.6 Range (a) and a close-up of one of the controllers (b). The icon and word "REAR" on the close-up of the control (b) are meant to inform the user of which burner is controlled by that control.

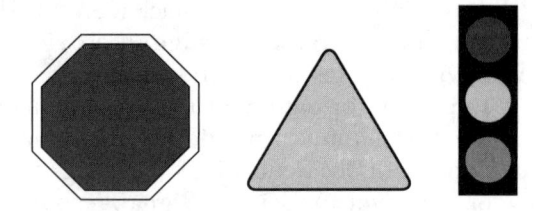

Figure 8.7 Three universally understood traffic symbols: the stop sign, yield sign, and traffic light scheme. (From Corel, "Corel Gallery: 10,000 Clipart Images," Ottawa, Canada: Corel Corporation, 1994, [9].)

Figure 8.8 Universal icons for audio and video recorders and players. These represent cultural and semantic conventions.

right-side control for the right burners. The burner control knob shown in Figure 8.6(b) is for the right rear burner, as indicated by the burner location icon. Hence, this device uses location as a substitutive representation, but also builds knowledge into the product via a burner location icon and textual information by each burner location.

Societal and cultural influences contribute to mappings in the form of universally or widely recognized signage, icons, and symbols. International travel, the World Wide Web, e-mail, global telecommunications, entertainment, and the global merchandising of consumer products help create this universal and global vocabulary. Figure 8.7 shows three universally understood traffic symbols. An international convention for using specific icons for controlling video and audio recorders and players has evolved. Figure 8.8 shows these conventions.

Other widely recognized symbols are shown in Figure 8.9. As with all mappings, a person's knowledge determines the informational value of these symbols. The linkage between the symbol and the meaning of the symbol is especially important if the mappings are meant to be constraints, such as the traffic symbols.

Figure 8.9 Other widely known symbols. The icons represent toilet facility for a disabled male, no smoking, female and male, and McDonald's Golden Arches. (From Corel, "Corel Gallery: 10,000 Clipart Images," Ottawa, Canada: Corel Corporation, 1994.)

Problems can arise from a design point of view in that cultural and semantic conventions are so ingrained that people typically are not aware that they are explicitly being employed in a particular design. This can lead to user errors, frustrations, or even accidents if users do not understand or share the cultural and semantic knowledge of the designers.

Constraints

> *Strategy 8.3: Use constraints so as to control the course of actions and prevent or reduce the possibility of the users doing the wrong thing*

Designers use constraints to build knowledge into the environment [5]. Physical constraints represent a typical design approach. For example, if a microwave oven door is opened while the oven is operating, the oven turns off. Similarly, the diesel fuel nozzle at gas stations is too large to fit into regular or non-diesel fuel tanks. The common three-pronged electric power plug and receptacle allow only one way to insert the plug. These examples illustrate physical constraints in the form of forcing functions. They are called forcing functions because they force users to behave in a certain way to avoid accidents or errors.

Cultural constraints control behavior based on societal conventions. In the United States people drive on the right side of the road, while in Great Britain people drive on the left side of the road. A universal set of driving conventions have evolved in terms of signs and signals, for example, the shape and color of stop signs, yield signs, and traffic lights. These societal conventions work as long as the users know and understand the meaning of the signs, shapes, and colors.

Further discussion regarding constraints appears in subsequent chapters.

Feedback

> *Strategy 8.4: Use feedback to keep the user informed as to the status of the entity's operations and the entity's response to user inputs*

Designers use feedback to inform users about the results or consequences of specific user actions. Users push a control panel button and feel a click or snap to indicate activation. Users press "return" on a computer keyboard and a symbol, picture, or text informs them of an action: the computer is loading a program, saving a file, making an Internet connection, or delivering e-mail. Such feedback lets users know that their input or control action has been received and is being acted upon. Without such feedback, users become frustrated and often make errors (pushing a button several times) because they do not believe that their intentions were received by the system as a result of their deliberate use of a controller.

Messages, and hopefully information, are provided by the use of warnings and feedback. Designers use a variety of signals, symbols, and icons to communicate with the user. Using colors, sounds, lights, smells, and tactile cues for warnings represents mappings (the association of a signal with a warning) and constraints in that the mappings are intended to control or limit behavior. For example, a buzzer that goes off when users forget their keys in the car's ignition or the flashing light and ringing bell that warns of an approaching train at a road/railroad crossing provides feedback. The smell of "natural" gas is an additive that provides warning of a gas leak and rumble strips on the side of roads to inform drivers that they have moved off the main roadway—both signals give users feedback as to the consequences of their actions.

Feedback also occurs while using a device or product; it informs the user as to the correctness of the user's conceptual model of product functioning. A funny example of this occurred at a restaurant. Patrons ordered beverages that were served in heavy, thick-glass mugs. The patrons were talking as the drinks were served so they did not notice how the server handled the mugs. A gentleman reached out, took hold of the mug, lifted it to drink, and got drenched. The man got drenched because his conceptual model was wrong. The mug was not what it appeared to be. Instead of being a heavy, thick-glass mug, it was lightweight plastic. Unfortunately, the gentleman's conceptual model, based on his observations and past experiences (affordance), had told his arm and hand muscles that this was a heavy mug, so he lifted accordingly.

The effectiveness of affordances, mappings, constraints, and their associated signals and symbols to inform depends on whether the user detects the signals, understands the signals, and can act on the information provided. Hence, the affordances, mappings, and constraints used to build knowledge into the environment are limited in their effectiveness in various ways. When a designer relies on "knowledge in the head" to interpret affordance, problems can arise if the user does not possess the required knowledge, lacks sufficient life experiences, or is cognitively impaired. Also, the user might, as with the "glass mug," misinterpret the affordances. The notion of affordance is simple in that it refers to the actual and perceived attributes of a product or process that suggest its uses. The notion is complex in that it includes things users have learned, such as cultural and semantic conventions and constraints, as well as a formal and informal (intuitive) knowledge of science (physics and chemistry) and mathematics.

Language

When communicating with a person via feedback, signage, or other techniques, language is a critical element. The person must be able to understand the language to comprehend the message. This leads to another common-sense strategy.

Strategy 9: Provide messages in a language and format that the people using the process or product will understand

This could mean using multiple languages, Braille, and/or sign language. It could also mean using icons, pictures, or symbols that are widely understood. It could mean using signals that are clearly understood, for example, sirens and flashing red light as warning signals. It could mean a combination of any of these.

Strategy 9.1: Use a language that users understand (English, Spanish, etc.)

The choice of language—that is, English, Spanish, sign language, Braille, etc.—is obvious. Information must be presented to users in a language they understand. It is common that electronic consumer equipment provide instruction manuals and materials in a variety of languages. Ford Motor Company's Rouge Plant, in Dearborn, Michigan, has signs in at least three languages: English, Arabic, and Spanish. Other plants have different combinations and numbers of languages, depending on the nationality mix of the workforce.

Strategy 9.2: Use an appropriate style of language—formal/informal, technical/nontechnical

Designers should use an appropriate level of language in the user's manual, operating instructions, technical references, and support material. It is bad design to write a users guide for the general population using highly technical, engineering-oriented descriptions and terms. One must match the technical level of the documentation to the intended audience. Likewise, one should not use complicated and convoluted prose in the documentation. A sign over a technical writer's desk sums it up: "Speak plainly, eschew neologisms!"

Strategy 9.3: Use an appropriate level (grade level) of language

As important as style is, the level of language used is equally important. A recent study on car seat instructions found that the typical instruction manual was written at the tenth-grade level [10]. Studies have shown that improper restraint use varies from 79 percent to 94 percent, an alarmingly high rate [10]. The major cause for this appears to be that the "instruction manuals are written at a reading level that exceeds the reading skills of most American consumers. These instructions should be rewritten at a lower reading level to encourage the proper installation of child safety seats" [10].

A 1992 study of adult literacy found that about 21 percent of the adult population over 16 years old (over 40 million people) could only read at or below the fifth-grade level. Twenty-five percent of adult Americans (about 50 million people) were classified as "marginally literate," reading at or below the eighth-grade level [11]. This is a poor commentary on the U.S.

educational system but the study results present a reality that designers must address.

Technically trained people also present a challenge. A study conducted by the Danfoss Company, a major Danish industrial manufacturer for refrigeration, heating, and motion controls, concluded that service manuals needed to address the language skills and differences for global users. For example, the Danfoss study concluded that the service manuals for American service technicians should be at a grade level lower than that for their European counterparts. The report goes on to say that "U.S. service techs are not so well educated, and are less trained vocationally" [12].

Strategy 9.4: Use universally or globally understood icons, symbols, or pictures for communications

Globalization of consumer products and ease of travel have facilitated the creation of universally accepted and standardized icons. Electronic equipment controls, traffic signage, signage for commonly used public facilities such as toilets, and automotive dashboard controls are examples of using universally understood icons and symbols. The globalization of consumer products and increased international travel has necessitated creating universally accepted icons and symbols for products and services. Additionally, people who cannot read and those who are cognitively impaired benefit from using universally accepted icons and symbols; these are, in essence, a simpler universal language.

Assuming that designers use the appropriate languages (including icons, symbols, or pictures), then the user must be able to comprehend the messages and take appropriate actions to use the product or process. Also, if the product or process' operating procedures are too complex, users will make errors, become frustrated, and avoid using the product or process. These lead to an additional design strategy.

Complexity

Strategy 10: Reduce the operational complexity of the entity

Designers need to design entities that are as simple to use as possible. While this seems obvious, it is often very difficult to realize. Cost, time, resources, and other constraints can influence the design process resulting in a product or process that meets cost or time constraints, but at the expense of complexity. After all, people are clever; they will adapt themselves to any quirk or idiosyncrasy of the product or process. While this generally is true, it also leads to errors, accidents, complaints, frustrations, and other negative consequences. Complexity is a difficult concept to formalize. From a design perspective, complexity will be considered to derive from task structure, memory requirements, and required reaction time.

Chapter eight: Cognitively sound

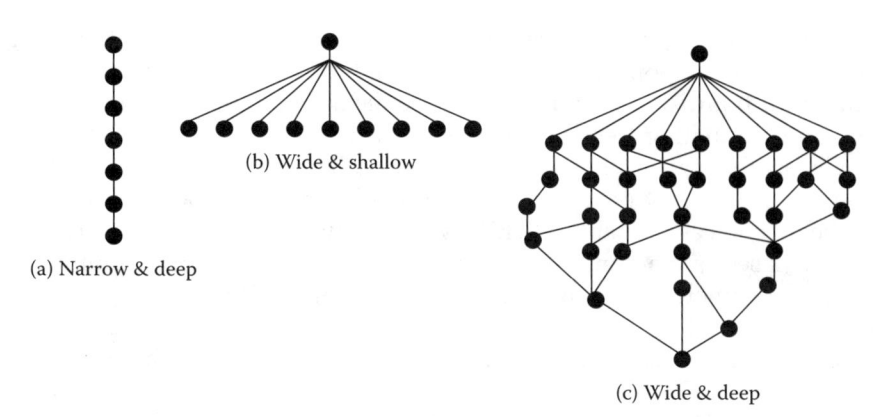

(a) Narrow & deep

(b) Wide & shallow

(c) Wide & deep

Figure 8.10 Three graphs illustrating different types of decision trees. (a) could represent the steps in a recipe; (b) could represent menu selections at a restaurant; and (c) could represent moves in a complex board game such as chess.

Strategy 10.1: Keep the task and activity structure as simple as possible
Graph theory provides a mathematical framework within which to analyze task structure complexity [13, 14]. Graph theory has been used to model decisions, relationships, traffic systems, computer networks, and variety of similar systems [14]. A graph consists of points and arcs [13]. From this deceptively simple base has developed a sophisticated branch of mathematics and engineering applications [13, 14].

Figure 8.10 shows three graphs representing three different types of decision or action tasks. Figure 8.10(a) is narrow and deep, in that there are many sequential actions but one essentially follows the other. This is typical of recipes or assembly instructions. Figure 8.10(b) is wide and shallow, in that there are many alternative actions possible, but once made not many more follow in sequence. Restaurant menus are typically wide and relatively shallow. Many selections exist for each course but there are relatively few courses. The third graph, Figure 8.10(c) is both wide and deep. This graph represents complex action or decision tasks, for example, the operating instructions for a nuclear plant or moves for a chess game. Most daily actions are typified by graphs with relatively narrow and/or shallow structures; people get up and decide what to wear, what to eat for breakfast. Work activities often require people to deal with very complex decision and action structures. Likewise, people often look for challenging leisure activities, for example, playing chess, rebuilding antique cars, or building experimental airplanes.

From a design perspective, engineers want to keep operating procedures as simple as possible. "Simple as possible" can be quantified by seeking decision trees that are relatively narrow and shallow, not a lot of choices at each decision level, and not many sequential steps or decisions. Graph theory provides a mathematical framework for quantifying decision trees and related structures such as flow diagrams, flowcharts, and network diagrams

[14]. Graph theory provides procedures for finding the shortest path through a graph from one point to another, the length of a path, and the number of paths from point "a" to point "b." These techniques are valuable for modeling and simulating the operating procedures of a process [15–17].

Graph theory provides a tool for visualizing decision trees and task structure. Hick's law provides a model relating human reaction time to the number of choices available. Hick's law basically says that a person's reaction time increases as the number of choices increases. Hick's law is another one of the psychometric laws, like Fitts' law, relating decision making (reaction time in this case) to the number of alternatives that need to be considered for selection by a motor response: pointing to, reaching, touching, etc.

Strategy 10.2: Design an entity so that the user's reaction time is within satisfactory bounds

Designers should not require people to quickly scan a number of choices and make decisions. Hick's law demonstrates that a person's reaction time increases as the number of choices increases [18, 19]. People need time to digest information and then formulate a response.

A classic example of Hick's law was the telephone operator's job with the old nondigital switching systems. An operator sat facing a board with up to a hundred plug-receptacles and had to connect an incoming line to the appropriate outgoing line. Call arrivals were random, as were the call routings. A more current example of Hick's law is a user interacting with a Web page on an Internet site [20, 21]. As the number of choices on the Web page increases, so does the time it takes for a user to make a choice. Figure 8.11 shows the "Web page from hell." Jay Boersma, a graphic artist from Chicago, Illinois, created this Web page as a model of what not to do, in terms of Web page design.

This Web page has so many choices and options that a user's reaction time is extremely compromised. Not only does this site ignore Hick's law, but it also ignores virtually all universal and accessible design principles. Scroll bars are everywhere and material is poorly blocked. Several sections that automatically scroll create attention conflicts because the user does not know just where to focus his or her attention. The "Web page from hell" is a very clever way to illustrate bad design strategies.

Conversely, the Google website is an example of good design (Figure 8.12). It is simple and straightforward. From a Hick's law perspective, it presents few and clearly distinguished choices. From a Fitts' law perspective, there is only one place to enter text and only a few relatively large buttons to click. There are no annoying banners or other graphics. The use of color is also limited.

Web designers are vitally concerned with the effectiveness of Web layouts for a variety of reasons, from marketing strategies to effective utilization online training and educational materials [20–22]. Most websites fall somewhere between the "Web page from hell" and the Google site with respect

Chapter eight: Cognitively sound 109

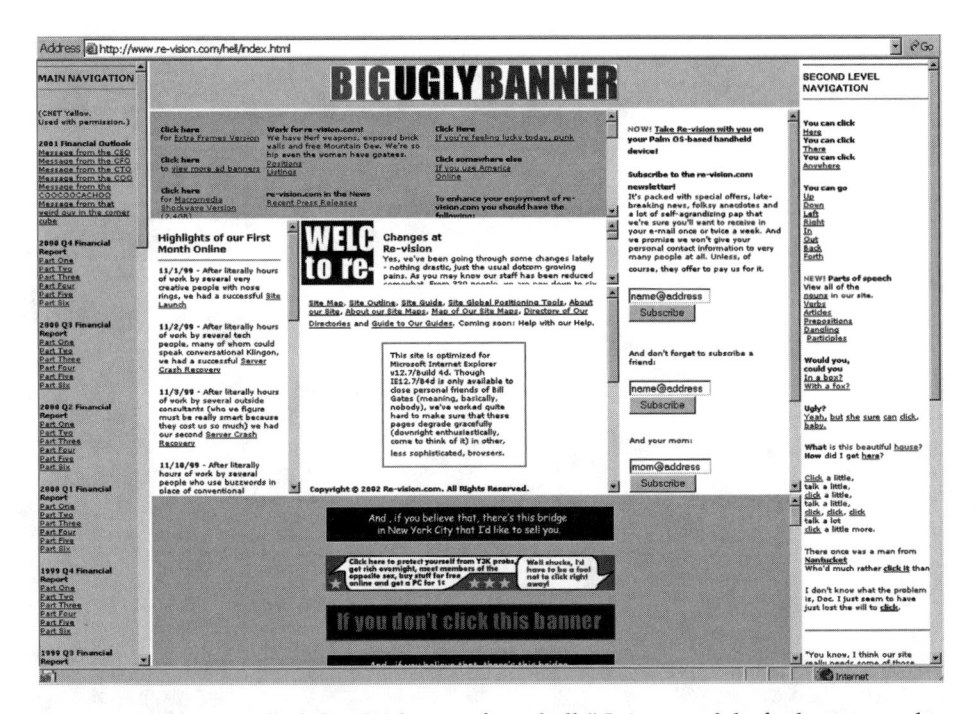

Figure 8.11 This is called the "Web page from hell." It is a model of what not to do. It was produced by Jay Boersma, a graphic artist working in Chicago, Illinois. (http://www.re-vision.com/hell/index.html.)

to good and bad design features. Web designers strive to distinguish links and use color, pictures, and standardized layouts to increase a user's discrimination of material and increase familiarity with the page layout. As noted, if the alternatives are easier to discriminate, the user's reaction time decreases. Also, as the user becomes familiar with the page layout, the reaction time becomes quicker.

Electronic devices exhibit features that invoke not only Hick's law, but also Fitt's law and ergonomic and perceptual issues. For example, consider small flip-open mobile phones (Figure 8.13). This design illustrates a number of design issues from ergonomics, perception, and both Fitt's and Hick's laws. The owner of this phone reported that her parents cannot use it because the control buttons are so small that they cannot operate them very well and they cannot easily read the display. They are not familiar with the control button layout, and all the choices are confusing. Even with glasses, the information layout on the display was too confusing for them to make sense out of it.

Figure 8.14 shows the control panel for a Bose Wave Radio/CD player. As with the universal remotes in Figure 7.1, the Bose device uses universal icons [(▲ increase, ▼ decrease) volume, (◄ backward, ► forward) for time, (► || play/pause, ■ stop)] to increase discrimination between various control choices. The icons are familiar symbols and an increase in discrimination

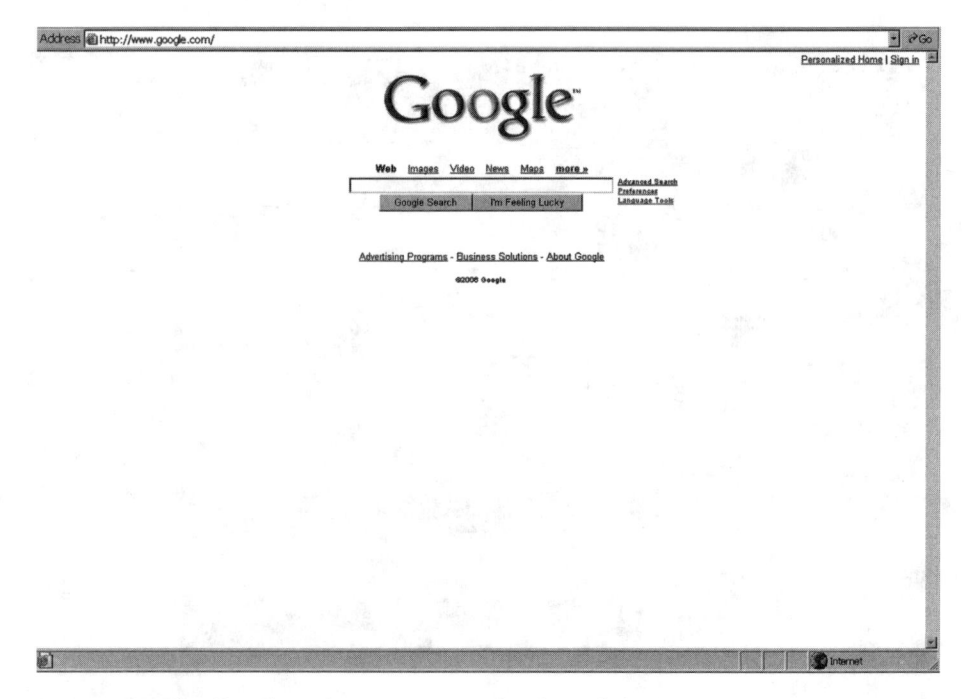

Figure 8.12 The Google website: an example of good design.

between choices tends to reduce a user's reaction time. As seen in Figure 8.14, the controls are grouped so that each area of the control panel is related to different, but specific functions. The grouping increases reaction time and reduces errors in that, for any given function, all the buttons are not equally likely choices; thus, users will focus their attention on the buttons that are associated with the desired function.

Task structure, graph theory, and Hick's law illustrate that as the number of choices or alternatives increases, the complexity of an entity increases. Complexity is also a function of memory requirements. Research has shown that people can remember about seven items in short-term memory [23]. Also, people do poorly at remembering details [23]. Designers should not require users to remember long sequences of actions, a large number of items from one Web page to another, or long sequences of arbitrary alphanumeric characters for PIN numbers, access codes, or passwords.

As complexity grows and memory requirements exceed seven items, people chunk items together and recode them into a hierarchy of categories that help them organize and remember increasingly complex amounts of information. Such memory-related human performance capabilities give rise to additional design strategies.

Strategy 10.3: Keep memory requirements simple

One approach to simplifying memory requirements is to design processes wherein long sequential decision tasks are not required; that is, avoid wide

Chapter eight: Cognitively sound 111

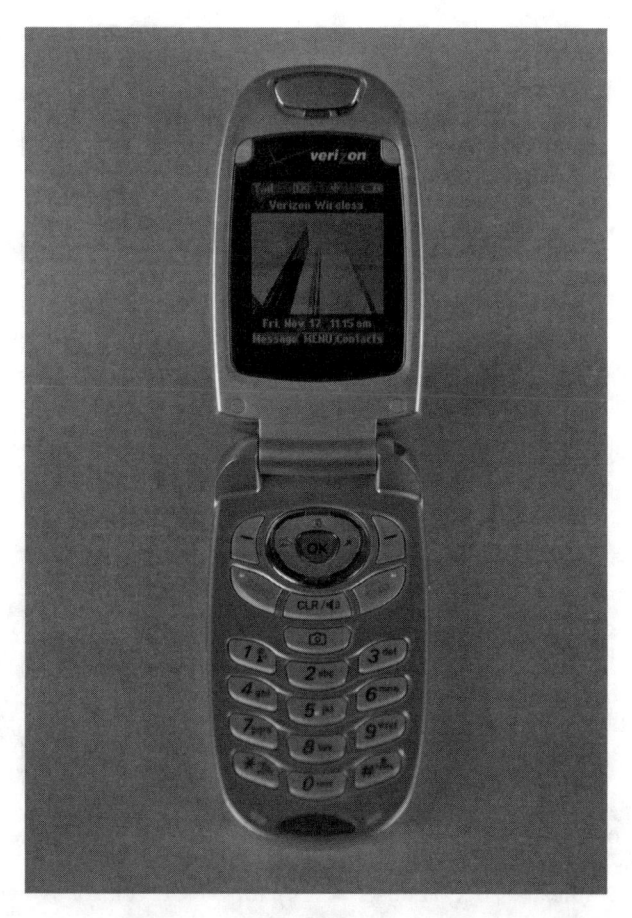

Figure 8.13 Typical small flip-open mobile phone. The keys on the keyboard are small, and the display is small, usually packed with a large amount of visual information.

and deep decision structures by designing a series of binary decision tasks. From a memory point of view, humans simply have trouble remembering a long sequence of actions that depend on the results of previous actions. People perform much better, and with fewer errors, if faced with a series of binary decisions [23].

Strategy 10.4: Avoid the use of modes

In terms of devices, *modes* refer to different modes of operation or functioning. For example, given the universal remote controller in Figure 7.1, the control buttons such as the numbers or up/down buttons produce different results, depending on the selected function (TV, VCR, CBL, AUX, PWR)—that is, mode of operation. A sports watch may have a time mode, an alarm mode, a stopwatch mode, and a pacing mode. The relatively few control buttons have different functions, depending on the functional mode of the watch. Typically, designers will resort to "modes of operation" when

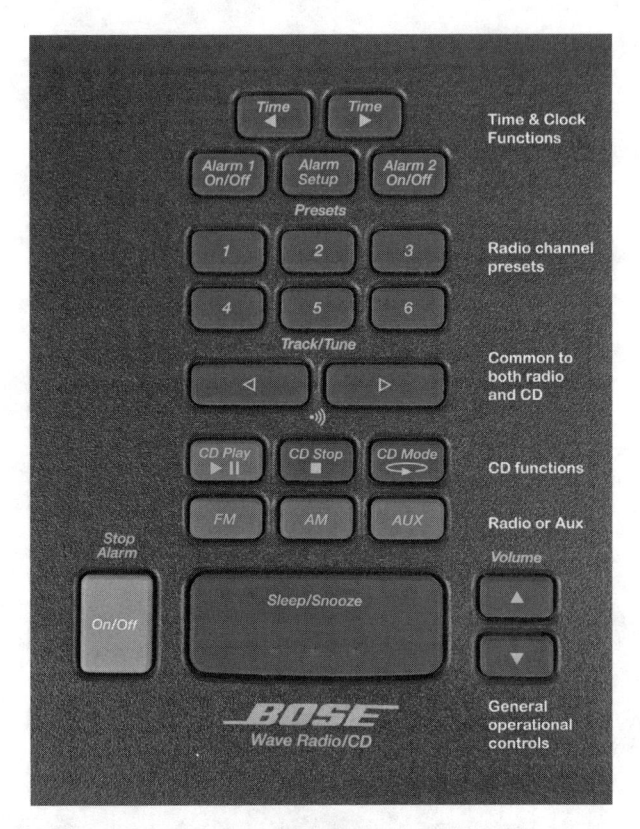

Figure 8.14 Control panel of a Bose Wave CD/Radio. The layout uses text to label the controls as well as universal icons (▲ increase, ▼ decrease) volume, (◄ backward, ► forward) for time, (► || play/pause, ■ stop) for the CD control. Note the grouping of controls according to function.

they have relatively few control buttons with which to work. Modes can be confusing because it may not be clear which mode the system is in and hence what button-pushing will do.

Note also that the universal remote (Figure 7.1) uses a spatial grouping of buttons. The mode selection buttons are across the top of the device. The number pad is in the middle. The channel up/down, volume up/down, play, stop, etc. functions are grouped together at the bottom of the device. Other miscellaneous functions are grouped toward the middle of the unit. Such a grouping of controls helps reduce the cognitive load of seeking and finding the proper control. The placement offers a search area for a type of control.

The Bose control panel (Figure 8.14) does not use a mode selection function. This means that each control button has the same function every time it is used. This greatly simplifies the cognitive demands of the task and hence a user's reaction time.

Chapter eight: Cognitively sound

Figure 8.15 The TI-84 calculator. The modifier keys select different modes of operations. Virtually every key has two functions depending on the mole.

The Texas Instruments calculator TI-84 is a standard among many high school programs and college students. Figure 8.15 shows the front face of the TI-84. This is a powerful tool that is relatively complex to understand and use and hence requires considerable training and experience before a person gains expertise.

Part of the TI-84's complexity, and power as a tool, derives from the many modes of operation it possesses. The modifier keys select modes and hence change the functions performed by most of the keyboard keys. This is a classic example of a device that supports a large number of different application programs with relatively few controls or keys. Hence, each key must take on different functions, depending on the mode of operation.

Education and training

Cognitive design strategies would not be complete if they did not address training and educational issues. People have different learning styles and different learning preferences. People are motivated differently. People have impairments that dictate different approaches to training and education. The Center for Applied Special Technology (CAST) has pioneered universal design for learning. The following strategies are drawn from CAST strategies.

Strategy 11: Use universal design for learning strategies developed by CAST for training or educational activities

People have different learning styles and preferences. Some people are visual learners and some are auditory learners. Some people learn best by reading and others by doing. Recent brain research has identified three interconnected neural networks that are specialized for different types of information process: (1) recognition networks for receiving and analyzing information, (that is, the what of learning); (2) strategic networks specialized for planning and action execution (that is, the how of learning); (3) affective networks specialized to set priorities and evaluate (that is, the why of learning) [24, Chap. 2]. If given a chance, people tend to select instructional modes, methods, and technologies that work best for them. From a design perspective, the challenge is to provide a choice of modes, methods, and technologies so that a broad spectrum of learning styles and preferences can be addressed.

Based on brain research and the associated learning theories, CAST has formulated three strategies for universal design for learning. These strategies give people choices.

Strategy 11.1: Provide the user with multiple means of engagement

Multiple means of engagement enable people with different interests, learning styles, or impairments to be more readily motivated to pursue and maintain engagement with the material (products, services, processes). More broadly speaking, designers must provide users with adjustability and choices as to how they interact and communicate with products, services, and processes.

Engagement occurs at different levels. From a sensory modality perspective, designers need to provide engagement for visual or auditory learners or people with visual or hearing impairments. Some people may need Braille, others a signer. From a methods perspective, engagement involves addressing those who want a factual or recognition-based approach, or those dealing with planning and taking actions. The emotional (affective) connections between a device, service, or process influence how readily people choose to use or avoid an entity. The emotional component derives from past experiences as well as promotional and societal influences.

From a technological perspective, designers need to provide different ways to access the entity. For example, computer access can be with a mouse, keyboard, touchscreen, and alternative keyboard devices such as head-pointers, eye-gaze monitors, or voice recognition systems.

Strategy 11.2: Provide the user with multiple means of representation
Multiple means of representing concepts, ideas, data, and information allow people with different learning styles or disabilities to more readily comprehend the essential concepts presented by the representations. Norman explicitly addresses representation issues in his book entitled *Things That Make Us Smart* [5]. The section on mappings touched on the importance of representations.

Some people want a written description of the material. Others prefer a picture or diagrammatic representation. Visual learners like concept maps, cause-effect diagrams, or flowcharts. Some people would like to see data in a table format, while others prefer using charts such as histograms, pie charts, bar charts, etc. The designer needs to consider the representation.

Strategy 11.3: Provide the user with multiple means of expression
Multiple means of expression provide people with a variety of ways to express themselves with respect to communicating, reporting, assessment, and evaluation. For example, creating a presentation using a written document but offering versions in large print, Braille, an electronic form readable by standard text-to-voice systems, or use of a signing avatar [25–27] offers multiple ways of expression. Other means of expression include visualization tools such as VISIO® [28] for the creation of flowcharts, network charts, cause-effect diagrams, timelines, and a variety of other representations. Concept mapping and visual organizational software are available, for example, Inspiration® [29] or MindJet® [30]. Of course, one may use video tapes, DVDs, CD/ROMs, computer-generated video files, or animation software. Such visual tools can be applied in education and collaborative activities such as planning and business process management.

Conclusions

§ 1193.41 of Section 255 of the Telecommunications Act of 1996 simply says that telecommunication system controls should be "operable with limited cognitive skills" providing "at least one mode that minimizes the cognitive, memory, language, and learning skills required of the user" [1]. The vagueness of this language speaks to the difficulty of design strategies related to cognitive issues. Yet the goal of cognitive soundness can be addressed by the systematic application of the design principles presented in this chapter.

Simplicity is a theme that keeps reemerging. Remember to "KISS," *Keep It Simple Stupid*. However, while we all know "simple" when we see it, simplicity is a deceptively complex concept. What is simple for one person is difficult for another. Simplicity depends on the demands of the task and the abilities of the user. Simplicity has many facets, as witnessed by the number of design strategies that address these different facets. Strategies 10.1 through 10.4 deal with task structure, the number of choices, memory requirements, and mode. Strategies 9.1 through 9.4 deal with language-related issues. Strategy 6 emphasizes the need for simple signaling structure and judgment techniques. Strategy 3 addresses the need for ergonomic simplicity. These strategies address fundamental human capabilities, strengths and weaknesses. The basic idea comes back to designing entities that have people doing what they, as humans, do best and having the designed entity complementing and supporting human capabilities.

The design process needs to create entities wherein the person and the designed entity work as partners to achieve desired objectives. This perspective suggests that focusing solely on human factors and functioning will not be sufficient to achieve the desired partnership in that the design must also consider the desired objectives of the task or process. The specific applications of the design strategies must reflect the broader process or task objectives. The following design principles—that is, flexibility, error-proofing, efficiency, and stability and predictability—focus on the process. As such, these principles tend to place constraints on the specific application of the three human factors principles.

References

1. Access Board, *Telecommunications Act Accessibility Guidelines: Final Rule*. Washington, D.C.: Federal Register, 36 CFR Part 1193, 1998.
2. T. Walsh, "Meijer Takes on Wal-Mart's Supercenters," in *Detroit Free Press*. Detroit, Saturday, November 6, 2004, pp. 14–15.
3. D. A. Norman, *The Design of Everyday Things*. Originally published in hardcover by Basic Books, 1988; Basic Books—Member of the Perseus Book Group, 2002.
4. D. J. Forrester, "Learning Theories," http://www.ucalgary.ca/~gnjantzi/learning_theories.htm, 2001.
5. D. A. Norman, *Things That Make Us Smart*. Cambridge, MA: Perseus Books, 1993.
6. R. F. Erlandson, "Universal and Accessible Design Principles," http://www.etl-lab.eng.wayne.edu/adrc/Definitions/UAD4.htm, 2003.
7. R. F. Erlandson, "Human Performance, Information Processing and Design," http://www.etl-lab.eng.wayne.edu/adrc/, 2003.
8. R. F. Erlandson, "Weber's Law, Fechner's Law, and Stevens' Power Law," http://www.etl-lab.eng.wayne.edu/adrc/, 2003.
9. Corel, "Corel Gallery: 10,000 Clipart Images," Ottawa, Canada: Corel Corporation, 1994.
10. M. V. Wegner and D. C. Girasek, "How Readable are Child Safety Seat Installation Instructions?" *Pediatrics*, 111, 588–591, 2003.
11. NCES, "National Adult Literacy Survey," Washington, D.C.: National Center for Education Statistics, 1992.

12. W. Sperschneider and N. T. Schmidt-Hansen, "Designing Future Scenarios for Electronic User Manuals," Danfoss Company, 2000.
13. F. Harary, *Graph Theory*, second printing. Reading, MA: Addison-Wesley Publishing, 1971.
14. M. N. S. Swamy and K. Thulasiraman, *Graphs, Networks, and Algorithms*. New York: John Wiley & Sons, 1981.
15. J. K. Pan and J. M. Tenenbaum, "An Intelligent Agent Framework for Enterprise Integration," in *Artificial Intelligence Applications in Manufacturing*, Fanili, Nau, and Kim, Eds. Cambridge, MA: MIT Press, 1992, pp. 349–383.
16. W. R. Miller, "The Definition of Design," http://www.tcdc.com/dphils/dphil1.htm, 1996.
17. V. Parunak, A. Ward, and J. Sauter, "A Systematic Market Approach to Distributed Constraint Problems," presented at Proceedings of the ICMAS'98, 1998.
18. W. E. Hick, "On the Rate of Gain of Information," *Q. J. Exp. Psychol.*, 4, 11–26, 1952.
19. R. Hyman, "Stimulus Information as a Determinant of Reaction Time," *J. Exp. Psychol.*, 45, 188–196, 1953.
20. K. Larson and M. Czerwinski, "Web Page Design: Implications of Memory, Structure and Scent for Information Retrieval," presented at CHI: Publication of the AMC, Los Angeles, CA, 1998.
21. P. Zaphiris and L. Mtei, "Depth vs Breadth in the Arrangement of Web Links," http://www.otal.umd.edu/SHORE/bs04/, 1998.
22. P. M. Sanderson, "Lecture: Information Theory: Hick's Law and Fitts' Law, PSY 258, IE 240, AVI 258: Human Factors in Human-Machine Systems," http://www.aviation.uiuc.edu/institute/acadprog/epjp/InfoTheory.html, 2000.
23. G. Miller, "The Magical Number Seven, Plus or Minus Two: Some Limits on Our Capacity for Processing Information," *Psycholog. Rev.*, 63, 81–97, 1956.
24. CAST, "Teaching Strategies," http://www.cast.org/strategies/, 1998.
25. I. Vcom3D, "Vcom3D, Inc., — Signing Software," http://www.vcom3d.com/, 2004.
26. M. Verlinden, C. Tijsseling, and H. Frowein, "A Signing Avatar on the WWW," presented at Gesture Workshop, 2001.
27. W3C-WAI, "Demonstration of Signing Avatar Technology as Used in the Signing Science Project," http://www.w3.org/WAI/RD/2004/06/sims-mov.htm, 2004.
28. Microsoft Corporation, "VISIO," http://office.microsoft.com/en-us/FX010857981033.aspx, 2005.
29. I. Inspiration Software, "Inspiration Software, Inc. – Homepage," http://www.inspiration.com/home.cfm, 2005.
30. Mindjet, "Mindjet Homepage," http://www.mindjet.com/us/, 2005.

chapter nine

Flexible

Chapter goals:

- State the flexible universal design principle and strategies.
- Provide examples that exemplify its process orientation.
- Discuss the relationship between flexibility and assistive technology.
- Explore the design trade-offs between flexibility and complexity.

Flexible

Principle

Design products, systems, and environments with enough flexibility so that they can be used and experienced by people of all abilities, to the greatest extent possible, without adaptations.

Discussion

Designing flexibility into an entity increases the likelihood that the products, systems, and environments can be used and experienced by people of all abilities, to the greatest extent possible, and without adaptations. However, increased flexibility does not come free, it can entail a more complicated design, increased complexity, slower performance, and increased cost. Flexibility typically results from providing the user multiple choices, and/or adjustability—for example, providing different input/output options when operating or using a product or process. Multiple modes of signaling and communicating when providing feedback and warnings ensure that people who cannot perceive a particular signal or communications channel will be able to perceive another mode.

Flexibility also derives from adjustability—for example, being able to adjust cell phone or car radio volume, or the television's color contrast. Automobiles provide adjustable steering wheels, seats, mirrors, dashboard light levels, and floor pedal placements (brakes and accelerator) and thereby make their products accessible to a much wider spectrum of people.

Note that these elements (that is, adjustable volume, adjustable seats, choice of mode of communication, etc.) were covered by strategies from the human factors group and as such focused on the more narrow person/product or person/system interaction. In this chapter the focus is on the process and a number of process attributes, such as logistics, material handling, storage, cueing, and scheduling. From a process design perspective, the designer should understand the market demands and customer preferences and abilities before implementing a specific design strategy. These process-level market and user-driven needs and preferences place constraints on how the specific human factors design strategies are implemented.

Ensuring that designed entities are compatible with assistive technologies (ATs) is another flexible design strategy. This is because, for many people with disabilities, assistive technology may be the only method of engagement with the product, service, or process. Hence, from a universal design perspective, providing a user the choice to access the designed entity with AT affords not only accessibility, but also flexibility. The fundamental strategy for flexibility is to provide choices.

Strategy 12: Provide the user with choices

Depending on the product, system, or process, users should have the opportunity to choose, for example, language, the mode of communication, the method of engagement, the scheduling of the engagement, the pace of the activity, and the frequency of the engagement. Consider two examples in relation to flexibility: (1) a home automation system and (2) a phone-based customer support system.

A home automation system can exhibit flexibility with respect to available services and access [1–4]. Functional, or process, flexibility derives from the constellation of services available via home automation. For example, one can control the scheduling and duration of operation of home appliances and equipment such as dishwashers, clothes washers and dryers, the heating/cooling cycles of furnaces and air conditioners, and lawn watering (time and sector). Home health care can be included in the broad category of home automation features. Advances in telecommunication technology allows for home health care using remote sensor technology to monitor a person's vital signs and notify emergency care providers or family members if problems arise [5]. In addition to providing everyone more options for everyday living, all of these services support independent living for the elderly and people with disabilities, thus providing all consumers with increased flexibility.

Functional, or process, requirements dictate how a user might physically access and communicate with the home automation system, that is, how a designer might utilize the human factors design strategies. If the functionality only demands access to the system within the home, then a home computer, simple keypad, or a switch-activated control might be sufficient.

Chapter nine: Flexible 121

If the functionality also demands remote access, then the system may have phone access or more sophisticated wireless remote control. For each of these forms of access, the functionality requirements may demand different communication techniques (from the human factors level) — for example, a keyboard, phone keypad, voice activation, or voice recognition.

Recall the discussion of Polanyi's systems theoretic principles regarding hierarchical systems (Chapter 5), wherein the higher level's functional requirements place constraints on the design options of the lower level. From this example it is clear that a system's functional requirements, from a process perspective, places constraints or demands on the application of the human factors design strategies.

Flexibility, for a phone-based customer support system, means how many people with as diverse native abilities as possible can access and successfully use the system for the intended purpose. There are many kinds of phone-based customer support systems, each with different functional requirements—for example, bus schedules or movie times at the local theater, or credit card and bank account information or technical support for computers or software. Phone-based systems offer flexibility in that they are typically available on demand, but they can be particularly troublesome for customers if long wait times or a tedious automated questioning process is encountered. Most phone systems provide access to touch-tone (frequency) and old-fashioned rotary dial (pulsed), legacy technology, telephones. However, some services have opted not to support legacy technology, and this can deny access to some people who do not have touch-tone (frequency) phones.

Most phone services offer the customer a choice of languages. Language options are also important for Web services and operating instructions for consumer products.

Strategy 12.1: Provide the user with a choice of language (English, French, etc.)

Allow users to select the language of choice. Global markets are forcing not only multinational corporations but everyone doing business globally to be responsive to the languages used to present a product or service. The European Union (EU) website, EUROPA [6], allows the user to select a version of the site in any of the EU member languages.

Electronic and information technology products typically come with printed operating instructions in at least three to four languages. A more high-tech approach is exemplified by Agilent Technologies. Agilent Technologies distributes a variety of electronic test equipment worldwide. Their 54600 benchtop portable oscilloscopes exemplify Agilent's product line with respect to building knowledge into the environment. The device has a Quick Help system that brings up a help message on the oscilloscope's display screen should the user request help. The system knows what the user is doing by monitoring the device's control settings and can provide a targeted message in 11 languages.

These examples illustrate one way in which globalization is driving universal design and how language options are dictated by different market and design requirements. EUROPA offers a language for each language of the EU members. Operating instructions for electronic products marketed in North America usually come in English, Spanish, and French, the three most frequent languages found in the United States, Mexico, and Canada. Agilent Technologies addresses its global market by providing 11 different languages. Again, the functional demands of the process place constraints on the choice of languages to include in the design.

Strategy 12.2: Provide the user with a choice of mode for communication

Allow users to select the mode of interaction: visual, auditory, Braille, sign language, etc. Television sets have the option of displaying captioning. Computers are starting to offer text-to-voice programs as part of their standard package. Another fascinating development is the use of signing avatars, that is, animated figures that provide sign language presentations of information [7]. Companies are marketing signing avatar programs for use at presentations, meetings, classrooms, and for online Web-based services [7]. The Enabling Technologies Laboratory provides a number of devices for the work environment that allows a user to select the mode of communication—visual or auditory [8–10]. There is a growing demand for hands-free mobile phone operation, especially in the car. The United Kingdom has legislation requiring hands-free mobile phone operations for drivers [11]. A number of companies are selling Bluetooth voice recognition kits for hands-free use [12].

The availability of affordable voice recognition technology is a real benefit to many physically and visually impaired individuals. Voice recognition technology renders many products and services accessible. In this sense, using voice recognition, when appropriate, provides benefits to both non-impaired (hands-free phone use while driving) and impaired users (access to phone service).

Return now to the flexibility of telecommunication systems and accessibility by individuals with disabilities. Can a hearing-impaired person access and use the system? Does the system support TTY devices? Can a person with no arms or limited use of his hands access and use the system by employing voice recognition technology? These questions imply that flexibility derives from compatibility with assistive technologies— that is, the entity must support a variety of assistive technologies that provide accessibility for individuals with impairments.

Strategy 12.3: Ensure that appropriate assistive technologies can be effectively used with the designed entity (that is, ensure compatibility with appropriate assistive technology)

For individuals with impairments, the assistive technology may be their only method of engagement with the product, process, or service. While universal design strives to reduce the need for assistive technology, it is unlikely that universal design will eliminate the need for assistive technology. Both Sections 255 (Telecommunications Act) and Section 508 (Rehabilitation Act) mandate that telecommunications products and services and electronic and information technology be compatible with appropriate assistive technology. Appropriate assistive technology means assistive technology that allows access to products and services for normal use. For example, a telecommunications device for voice communications must be compatible with TTY devices, but not necessarily with a joystick wheelchair control device. The TTY is appropriate for the telecommunications system but the joystick interface is not.

Ensuring compatibility seems fairly straightforward; but as seen with the digital cell phone/hearing aid interface issues discussed in Chapter 4 (Compatibility section), the requirement has raised issues as to who is responsible for ensuring such compatibility—the designer and manufacturer of products and services or the designer and manufacturer of the assistive technology? The issue with respect to digital cell phones and hearing aids comes down to who must change their design so as to avoid interference. As noted, ANSI C63.19 measurement standards lay out a standardized method for assessing hearing aid/cell phone compatibility, but the standards do not address how this compatibility should be achieved. Each side would like the other to design a product that deals with the interference problem [13]. In this case, solutions to the interference problems are emerging in the form of new cell phone antenna designs [14] and external devices such as the Motorola Hands-Free Neckloop accessory [15].

The Assistive Technology Industry Association (ATIA) is a trade association of the manufacturers and suppliers of assistive technology. This organization is working hard within the business community to address legally mandated compatibility requirements as well as the longer-term issues of compatibility across all products and services, whether covered by legal mandates or not [16].

Strategy 13: Provide adjustability and mobility

Choices need to extend to how a person interacts with the product or process. Such flexibility goes back to the earlier strategies for ergonomics, perception, and reaction times. Because flexibility is viewed as a process dimension, the process demands place constraints on how the ergonomic and perceptual design strategies are implemented.

Strategy 13.1: Provide ergonomic and environmental adjustability
Allow users to adjust the entity to their individual environmental preferences and body size and shape. Cars and trucks provide adjustable seats, steering wheels, floor pedal positioning, and mirror positions. Additionally, one can adjust the internal temperature, and in some models the temperature can be set for each side of the car as well as front seats and back seats. One can control the lights for different areas of the car. One can control the distribution

and intensity of music in the car. In some models, the driver can select either a digital display or an analog display for speed, temperature, pressure, etc.

Evolving housing needs exemplify market-driven universal design in that the combination of more flexible physical floorplans and home automation increases flexibility. Flexibility in housing comes from providing multiple options for use of space and an infrastructure that supports quick and easy accommodations. Linda Reimer, president of Design Basics, a home plan design company, notes that baby boomers in their 40s and 50s are becoming empty nesters, and "more and more of them are choosing flexible homes that will allow them to age in place" [17]. Flexibility in housing can be accomplished by creating fewer rooms that are larger and that can be used for different purposes—for example, a room that can serve as a den, living room, or guest room. One can anticipate change and plan for it in the design. For example, build wheelchair-accessible bathrooms and large walk-in showers. The larger bathrooms are spacious and the larger showers could easily accommodate a seat, if necessary. Likewise, special supports should be added behind the walls during construction where grab bars might be installed should the need arise.

Home automation allows for the customization of the living environment, the distribution of light, music, temperature control, the scheduling of clothes and dishwashing cycles, lawn watering, and home security features [2–4, 18]. Flexible layouts and home automation address broad market segments, simplifying life and providing creature comforts for everyone while allowing people with impairments to remain as independent as possible, thereby reducing the disabling pressures of the environment. Home automation also exemplifies the linkage between technological development and society's perception of disability. If people can live independently in their homes using technology, then they do not see themselves as disabled. Furthermore, society does not see them as disabled.

Strategy 13.2: Provide perceptual adjustability

This strategy refers back to Strategy 5 in Chapter 7. The designed entity should allow users to adjust signals for their own sensory needs. Volume control for a public kiosk has a different implementation than volume control for a home entertainment system or volume control on an industrial control system. While fundamentally addressing the same human factors issue, the environment and user needs vary greatly and hence, each specific design strategy will have a different expression in implementation.

Strategy 13.3: Provide adjustable response times

If possible and reasonable, allow users to determine their own response times to product or process queries and control functions. For example, *Sticky Keys* provide a way to perform simultaneous key operations sequentially rather than together; for example, instead of entering <Ctrl>z together, the *Sticky Key* option allows you to enter first <Ctrl> then z for the same results. People who can only type with one hand, or use a mouth-stick, head pointer, or eye-gaze system for keyboard access find this a valuable option. Also, computer users can adjust the "double-click speed" of a mouse. Three areas of Section 508 deal with user response times: (1) § 1194.22(p) deals with Web-based intranet and Internet information and application systems; (2) § 1194.23(d) deals with telephone-based services; and (3) § 1194.25(b) deals with self-contained and closed products. All state that when a timed response is required, the user must be alerted and given sufficient time to indicate if more time is required [19, 20].

One can generalize process response times to the workplace and the interface between the worker and the job process. Flexibility with respect to work processes arises from many sources: times for starting and finishing a job, tempo of the work, the sequencing of different parts of the job, conditions for participating in the work activity, time and where the work will occur, technology for participating in various aspects of the job, delivery channels for inventory, and materials and supplies. Workplace flexibility also derives from the use of agile system design strategies.

Strategy 13.4: Use agile system design strategies—adjustable and mobile

Agile manufacturing refers to the integration of organization, people, and technology into a coordinated whole [21]. The application of *agile systems concepts and strategies* promote flexibility. *Agile* refers to system properties that are (1) *simple*: simple design, simple tools, and simple assembly; (2) *mobile*: portable, lightweight devices; (3) *adjustable*: devices can quickly and easily reflect changes in the work environment, the product, and the product demand; and (4) *reusable*: materials can be reused as products, processes, or assembly volumes change [22, 23]. The life of the device can be geared to the life of the job, and then workers can reuse the components in a new device or system configuration. Agile devices directly address ergonomic issues such as material handling, storage, and parts presentation/positioning for the workers.

Pipe and joint technology is a classic example of an agile system technology, and Creform® Corporation is one of the largest companies manufacturing such technology. Creform® devices are found throughout the manufacturing and assembly plants of all the major automotive companies.

Figure 9.1 shows two dramatically different applications of the pipe and joint technology. Figure 9.1(a) and (b) show a Creform® flow-rack workstation in an automotive assembly plant. Figure 9.1(c) shows a Creform® beanbag chair made by a team of special education teachers and occupational therapists at an Enabling Technologies Laboratory (ETL), the Agile Device Workshop. A mobile Creform® frame was designed to hold a beanbag, which then provided positioning support for a severely physically impaired child. The support allowed the child to use a switch to interact with a computer. Over 200 unique devices have been designed and built by teachers and

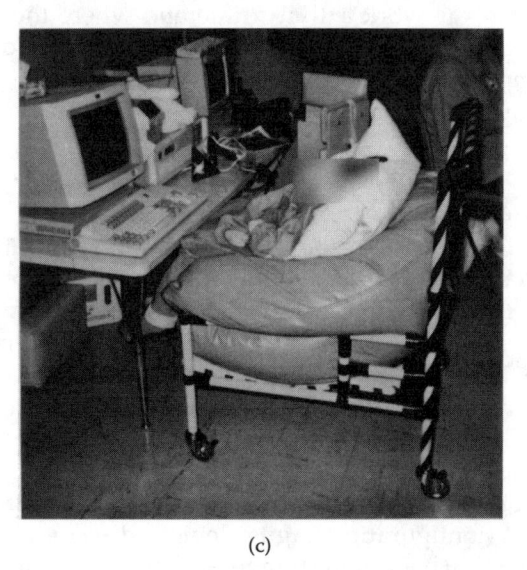

Figure 9.1 (a) and (b) show industrial applications of the Creform® technology, and (c) shows a mobile beanbag chair frame being used to support a severely physically impaired child so as to access a computer.

therapists at the ETL workshops. These agile devices are being used in schools throughout southeastern Michigan [10].

Strategy 14: Build flexibility into service delivery systems and work processes

Services provided to consumers must provide flexibility along several dimensions. Table 9.1 presents some dimensions of flexibility as related to a service offering. The major dimensions are time, content, requirements, delivery, and logistics. The designer and provider can elect to be flexible or

Table 9.1 Four Key Dimensions of Service Flexibility

Dimension	Fixed ← Dimension → Flexible
Time	Times for starting and finishing a service
	Times for submitting transactions and interacting with the service provider
	Tempo/pace of conducting transactions
Content	Content, type, and quality of service provided
	Sequence of transactions and services provided
	Orientation of the service (sales, information)
	Instructional material, users manuals, guidelines, and procedures
Requirements	Conditions for participation
Delivery and logistics	Time and place where contact with service provider occurs
	Methods and technology for obtaining support and making contact
	Types of help, communication available, and technology required
	Location and technology for participating in various aspects of service delivery
	Delivery channels for information, content, and communication

Source: Adapted from B. Collis and J. Moonen, *Flexible Learning in a Digital World*. London, U.K.: Kogan Page, 2001.

fixed with respect to a variety of elements associated with each major dimension. Table 9.1 modifies flexibility dimensions for the delivery of educational services as presented by Collis and Moonen [24].

There should be flexibility built into the workplace and the interface between the worker and the job process. This flexibility arises from many sources: times for starting and finishing a job, tempo of the work, the sequencing of different parts of the job, conditions for participating in the work activity, time and where the work will occur, technology for participating in various aspects of the job, delivery channels for inventory, and materials and supplies. The *administrative controls* cited in the OSHA ergonomics standards document speak to such flexibility.

In addition to the elements presented in Table 9.1, choices in training and educational activities derive from Strategy 11. This strategy is repeated here to emphasize its importance in providing flexibility.

> Strategy 14.1: *Provide the user with choices for educational or training activities consistent with the CAST universal design for learning strategies presented in strategy 11*

The CAST universal design for learning strategies addresses educational services and processes. Providing options for language, communications, and for the three multiples (engagement, representation, and expression) generalizes to all products and processes and includes service delivery systems and job processes.

A design challenge

Designing flexibility into an entity creates a more complex and demanding design process. The resultant designed entity gains parts, linkages, and interconnections, adding cost and complexity to fabrication and implementation. Another design challenge associated with enhanced flexibility is the need to provide affordances, mappings, and feedback so as to inform the user that there are options and/or that things can be adjusted, moved, or repositioned. Because flexibility often adds complexity to the designed entity, designers are forced to carefully think about building knowledge into their design, that is, additional affordances, mappings, constraints, and feedback.

The examples for Strategy 12.1 dealing with providing language options illustrate the design challenge: what languages and how many languages do we utilize? Newer HDTVs offer a large number of options for the consumer. These options are typically presented to the user by way of on-screen menus that guide the user through the choices. A detailed instruction manual accompanies the on-screen menus. The manuals come in different languages and the on-screen menu allows the user to select a corresponding language for on-screen menu instructions.

The HDTVs also offer a variety of human factor options, some for accessibility and some for preference. Accessibility options include the standard volume, brightness, and contrast controls, as well as closed captioning. Preference options include screen format from full, wide, zoom, and zoom-wide. From a designer's perspective, it is a challenge to communicate all these options to the consumer. While the options do add flexibility, they can also lead to errors when operating the HDTV.

Conclusions

Flexibility increases the likelihood that the designed entity will be used and experienced by the greatest number of people to the greatest extent possible, without adaptations. Providing choices and adjustability (including compatibility with assistive technology) are the most fundamental strategies for achieving flexibility. The nature and number of the choices and adjustability features depend on the process, objectives, and costs.

Flexibility can be considered a double-edged sword. On the one hand, flexibility can enhance human functioning and performance from the ergonomic, perceptual, and cognitive perspective. This being the case, people can select features and options that make them less error-prone and more efficient. If a person is less error-prone and more efficient at performing a task or while engaged in a process, the process will most likely run more consistently and smoothly, with fewer slow-downs and fewer accidents.

On the other hand, increased flexibility can introduce design overhead into the design process and introduce increased complexity into the designed entity. The presentation of options and feature selection must be managed.

If not done well, the increased complexity will outweigh any benefits brought on by the increased flexibility.

Flexibility also comes when products and processes are compatible with assistive technology in that people with a broader spectrum of functional abilities can access and use the products and services.

References

1. Future Computing Environments, Future Computing Environments Smart House Web Survey, http://www.cc.gatech.edu/fce/seminar/fa98-info/smart_homes.html, 1998.
2. HAI, Home Automation, Inc Website, http://www.homeauto.com/, 2004.
3. E. D. Mynatt, The Aware Home, http://www.cc.gatech.edu/fce/ahri/, 2000.
4. The Home Automation Forum, The Home Automation Forum Website, http://www.homeautomationforum.com/, 2004.
5. J. Winters and W. Herman, *Report of the Workshop on Home Care Technologies for the 21st Century*. Rockville, MD: The Catholic University of America, 1999.
6. EUROPA, EUROPA: Gateway to the European Union, http://europa.eu/index_en.htm, 2006.
7. I. Vcom3D, Vcom3D, Inc., Signing Software, http://www.vcom3d.com/, 2004.
8. R. F. Erlandson and D. Sant, "Poka-Yoke Process Controller Designed for Individuals with Cognitive Impairments," *Assistive Technol.*, 10, 102–112, 1998.
9. R. F. Erlandson, M. Noblet, and J. Phelps, "Impact of Poka-Yoke Device on Job Performance of Individuals with Cognitive Impairments," *IEEE Trans. Rehabilitation Engineering*, 6, 269–276, 1998.
10. ETL, Enabling Technologies Laboratory: Home Page, http://webpages.eng.wayne.edu/etl/, 2000.
11. P. Newswire, Hands-Off! New Research Shows Mobile Phone Legislation Connecting with Drivers, http://www.findarticles.com/p/articles/mi_m4PRN/is_2004_April_5/ai_114909820, 2004.
12. cellPoint, Parrot CK3000™ Bluetooth Voice Recognition Car Kit, http://shop.store.yahoo.com/cellpoint/parckbluetvo.html, 2004.
13. ANSI, ANSI C63.19: American National Standard for Methods of Measurement of Compatibility between Wireless Communication Devices and Hearing Aids, ANSI 2001.
14. RFSafe.Com, RF Safe Approved Cell Phones Are ADA & FCC Compliant Already, http://www.rfsafe.com/an_rf_safe_approved_phone_is_ada.htm, 2004.
15. Motorola. Inc., Hands-Free Neckloop Accessory, http://commerce.motorola.com/consumer/QWhtml/accessibility/prodservices.html, 2004.
16. ATIA, Assistive Technology Industry Association Web Site, http://www.atia.org/, 2004.
17. L. Reimer, New Homes for Aging Boomers, http://www.designbasics.com/homebuyers/designtrends-10.asp?FormUse=homebuyers, 2003.
18. Smarthome, SMARTHOME Website Homepage, http://www.smarthome.com/index.html, 2004.

19. Access Board, Guide to the Section 508 Standards for Electronic and Information Technology, http://www.access-board.gov/sec508/guide/index.htm, 2004.
20. Access Board, *Electronic and Information Technology Accessibility Standards*. Washington, D.C.: *Federal Register*, 36 CFR Part 1194, 2000.
21. J. Ferris, Agile Manufacturing – lecture notes, file:///C:/Undergradute%20Book/Agile%20Design/256,1,Agile Manufacturing, 2002.
22. CREFORM Corporation, The CREFORM System, *CREFORM Material Handling Systems*, 2002.
23. J. D. Kasarda and D. A. Rondinelli, Innovative Infrastructure for Agile Manufacturers, *MIT Sloan Management Review*, Vol. 38, 73–82, 1998.
24. B. Collis and J. Moonen, *Flexible Learning in a Digital World*. London, U.K.: Kogan, 2001.

chapter ten

Error-managed (proofed)

Chapter goals:

- State the error-management/error-proofing universal design principle.
- Provide design strategies that address the error-management/error-proofing universal design principle.
- Provide examples of applying these design strategies showing how the process-level functional requirements place constraints on the use of forcing functions, feedback, and constraints.

Error-managed (error-proofed)

Principle

Entities must be designed so that they support doing the right thing. It is important to create a design wherein errors can be managed.

Discussion

To properly design an error-management or error-proofing strategy, it is essential to understand the functions of and reason for designing a product, system, or process. Designers strive to eliminate errors when consumers use their products and when workers manufacture and assemble products in a production system. However, in an educational or training system, where people learn by making mistakes, one should error-proof those elements of the process not critical to the learning objectives, and then manage the errors in the process segment associated with the educational objectives so as to facilitate the educational process. However, even in such educational and training settings, it is important to error-proof activities and processes not directly related to the educational objectives. Hence, this discussion focuses on error-proofing strategies and techniques.

Error-proofing is a fundamental *kaizen* (continuous improvement) strategy [1, 2]. *Poka-yoke*, a Japanese term for error-proofing, accepts that human beings are forgetful and tend to make mistakes. Shimbum, a Japanese proponent of *poka-yoke*, states that "too often we blame people for making mistakes. Especially in the workplace, this attitude not only discourages workers and lowers morale, but it does not solve the problem. *Poka-yoke* is a technique for avoiding simple human error at work" [3].

It is important to make a distinction between a *poka-yoke* intervention and an accommodation, in that accommodations are specifically mentioned as a component of accessible design [4, 5]. Accommodations represent a very narrow approach to productivity enhancement and are targeted at the worker with a disability. On the other hand, a *poka-yoke* intervention is designed to improve overall process productivity and quality for all workers, thus making it a more universal strategy.

An intervention for a worker with a disability may or may not be a *poka-yoke* intervention. If the intervention reduces error and improves the performance for everyone doing the task, it is a *poka-yoke* intervention. The fuel clamp assembly fixture described in Chapter 4 is a *poka-yoke* intervention because it improved performance and quality for all workers. If the intervention would slow down or hinder a worker who does not have a disability, it is an accommodation. For example, a specialized fixture designed to hold open the clamps on a pants hanger used in clothing stores so that an individual with the use of only one hand can perform the task of placing pants on a hanger is an accommodation. This device hinders workers having the use of two hands.

The pressures of lean production and higher quality standards imposed by the QS (Quality System) and ISO (International Standards Organization) certification requirements are accelerating the introduction of error-proofing strategies into assembly, manufacturing, office, and service jobs [1, 6]. High-quality products and services are fundamental to lean production. Ideally, one strives for zero defects in products or services [2]. The changing demographics of the workforce, cross-training, and job rotation require that work processes and jobs provide more feedback and support to the workers to not only reduce error, but also speed up on-the-job learning.

QS and ISO certification requires implementing quality improvement strategies; error-proofing is a core element of such strategies. Relatively inexpensive electronics and smart sensors with embedded microcomputers can be integrated into assembly, packaging, sorting, and counting tasks to provide feedback to the workers [7]. Color coding, icons, and other visual controls can be used as error-reducing tools in office and service jobs [3, 8, 9].

Error-proofing is a core strategy for achieving the objectives associated with lean production and quality control programs. For workers without disabilities, error-proofing strategies fight boredom, fatigue, and other distractions. For individuals with disabilities, error-proofing strategies provide an essential dialogue between the job and person that enable the person to actually perform the job.

Design strategies

In his book *The Design of Everyday Things* [10], Norman divides errors into two categories: (1) slips and (2) mistakes. Slips are due to inattention, boredom, fatigue, and reacting too quickly. Mistakes are thoughtful in that one digests information or data and then does the wrong thing. Mistakes are made if the information or data are incorrect, vague, or misleading, or the user perceives something incorrectly and thereby reaches the wrong conclusion. From a design point of view, these sources of errors present different challenges.

Slips

The first category of errors to consider is slips—errors due to inattention, forgetfulness, boredom, or fatigue [10].

Strategy 15: Use a three-staged approach to error-proofing

Shimbun presents this three-staged strategy for error-proofing in his book *Poka-yoke: Improving Product Quality by Preventing Defects* [3]:

1. Prevent errors at the source.
2. Provide a warning that an error has or is about to occur.
3. Provide quick and easy recovery if an error has occurred.

These are process-oriented strategies. Application of these process strategies places constraints on how a designer applies the human factors level principles and strategies.

Strategy 15.1: Prevent errors at the source

The most effective approach to error-proofing is to design an entity so that it is impossible or very difficult to make an error. Constraints called forcing functions are the most effective way to implement this approach [10]. Designers can implement forcing functions in a number of ways. One is to use direct physical constraints, that is, some physical property that constrains an action. For example, the diameter of a leaded gas tank hose nozzle is too large to fit in a gas tank receptacle for a car requiring unleaded gas. A three-pronged electrical power cord connector will only go into the wall outlet one way. Tall multi-drawer file cabinets often have a feature that allows only one drawer at a time to open, so as to avoid tipping the file cabinet. USB and other computer connectors possess grooves, shaped frames, notches, and other physical features that only allow them to join with a complementing connector. Figure 10.1 illustrates three different cables and connectors embodying three different forcing functions provided by the physical constraints of the connectors.

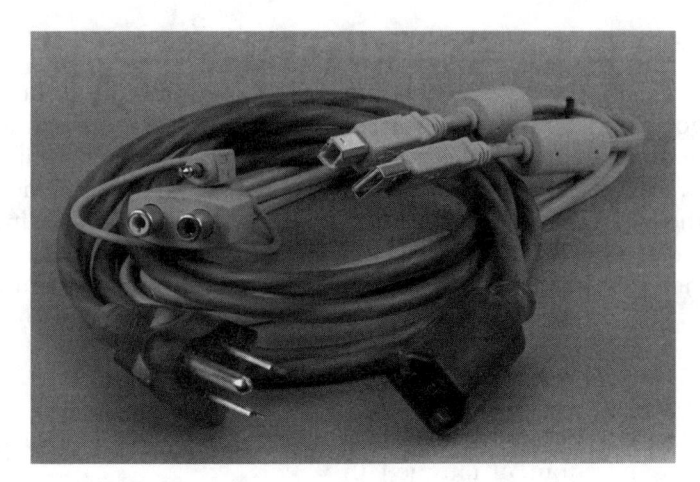

Figure 10.1 A variety of connectors: three-pronged power cord, USB cable and connectors, and a stereo audio connector. Each connector has a unique physical configuration ensuring only one way to mate the connectors. These connectors exemplify forcing functions.

Another forcing function example is an electric fuse. In this case, a metal circuit element heats up due to current flow. If the current exceeds a specified amount, the metal element melts and breaks apart, causing the electric current to cease. In this example, a physical property of the sensing material is used to bring about the error-proofing action.

The shut-off valves on gas pump nozzles typically use the physics of the venturi effect to control a mechanical switch. There is a small hole near the tip of the nozzle with a small pipe leading back from the hole to the handle. As gas fills the tank, it pulls air through the pipe. When the tank is full, the pipe tip near the end of the nozzle is submerged in gas. Now, much more suction is needed and a vacuum develops in the middle of the pipe. The vacuum can be used to flip a lever that cuts off the gas [11]. This is another example of a forcing function but the forcing action is based on the physics of the venturi effect.

Another error-proofing technique is to use a sensor to detect an error or potentially dangerous situation and then communicate this information to a process controller that can stop operations. For example, a microwave oven is designed so that the microwave energy turns off if the door is opened. An electronic or mechanical sensor senses that the microwave oven door has been opened and the sensor's change of state causes the system to turn off the microwave power. Likewise, a car's key will not come out of the ignition unless the gear shift is in the "park" position; either a mechanical linkage or electronic sensing mechanism linked to the gear shift keeps the ignition key in place until the gear shift is set to the "park" position.

In all these cases, the system's design prevents, or makes it very difficult, for the user to make an error because designers have used "at the source" error-proofing.

Strategy 15.2: Provide a warning that an error has or is about to occur
If it is not possible to prevent the error at the source, then design the system so that the user is warned that an error has occurred. For example, a warning signal goes off if users do not buckle their seatbelts in a vehicle. A laptop computer system provides a warning message when its battery is running low. Cell phones beep or signal in some way when their power is running low. In all these cases, users are being warned that an error has or is about to occur. Users can heed the warning and avoid the consequences of the error, or they can ignore the warning and suffer the consequences of the error.

The following two work-related examples exemplify error-proofing using a warning or indicator. They also illustrate how universal design can redefine the "essential functions" of a job and thereby render the job more accessible under Title I of the ADA. The first example is the packaging of brake fittings and comes from Shingo [2]. Before the improvement there were occasional counting errors in the packaging of the brake fittings. The fittings were all manufactured to tight specifications and therefore had uniform dimensions. The improvement took advantage of this physical property of the fittings. A jig was built to stack the fittings. A notch marked the 150-count height. Figure 10.2 diagrammatically illustrates the idea. Shingo noted there were no errors after this *poka-yoke* improvement.

The notch was an indicator (warning) that the worker had reached 150 brake fittings. Counting the brake fittings is much like counting coins. Figure 10.3 shows a commercially available collection of coin counting fixtures, for

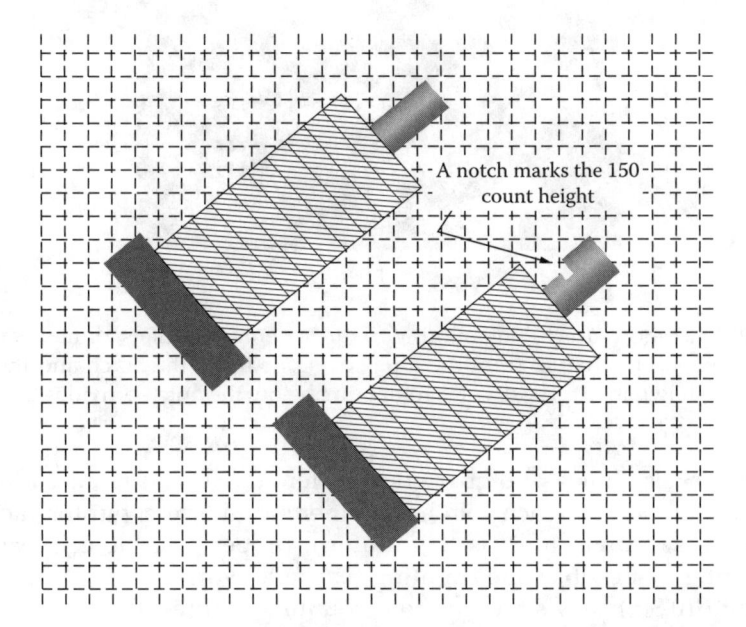

Figure 10.2 A *poka-yoke* intervention to eliminate errors in counting brake fittings. The method works because of the uniform dimensions of the brake fittings.

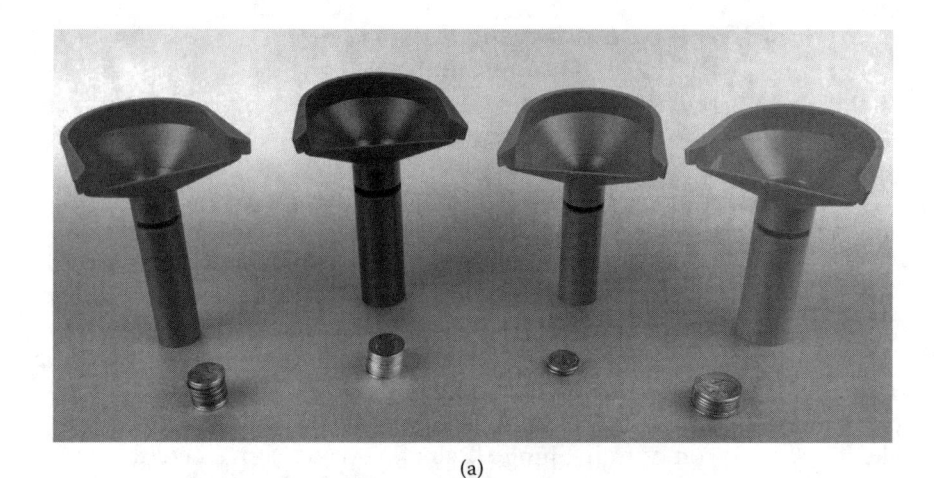

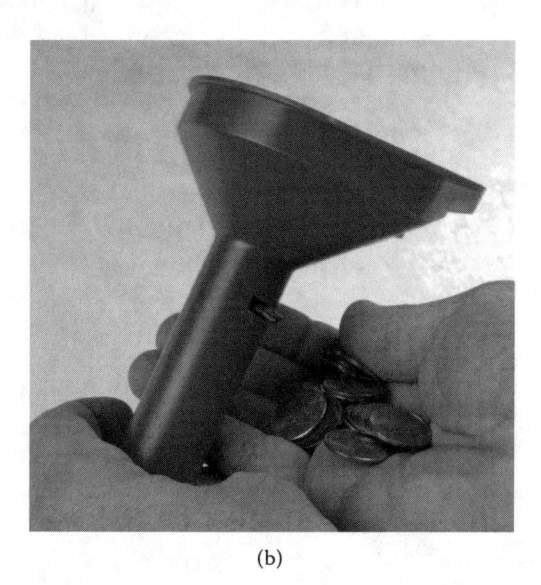

Figure 10.3 (a) shows four coin-counting fixtures: pennies, nickels, dimes, and quarters. One fills up the fixture from the top. Coins pile up to the notch and then they slide out as shown in (b). Those coins remaining below the notch provides the correct count.

pennies, nickels, dimes, and quarters. A different fixture is required for each denomination because each coin has a different thickness, but for each coin the thickness is uniform. It is this uniform property of the coins and the brake fitting that enables the counting fixture to work.

Figure 10.3(a) shows all four coin counting fixtures. Each fixture has a notch cut into the tube at a height that corresponds to the desired count, that is, the number of coins required for a bank roll of the specific coin (e.g.,

50 for pennies). Figure 10.3(b) shows how this works. Pennies are falling out of the fixture at the notch level, leaving the correct count behind.

The brake-fitting fixture is a process redesign with the specific goal of error-proofing. The goal of error-proofing this specific task places constraints on the human factors principles. The fixture *builds knowledge into the process* via the *feedback* provided by the notch. The process took advantage of a physical property of the brake fitting. One could have placed a notch at a height of 10 fittings and had the worker collect 15 of these "10 counts," but that is more complicated (relying on short-term memory), and hence more error-prone than just one notch at the 150 mark. Hence the process goal of error-proofing constrained the use the human factors principles *building knowledge into the process* and *feedback* by also requiring the *least complicated* process, in terms of *memory requirements*. In so doing, one produced a very simple and less error-prone process.

With respect to Title I of the ADA, the *essential function* of the "before" method was being able to count up to 150. This is a relatively error-prone task for anybody. The new design catches the error at the source; it uses an indicator or warning, the notch. The essential function of the new method is pattern recognition, that is, being able to understand what the notch signifies. The new method significantly simplifies the cognitive demands of the task and renders it accessible to workers who cannot count but who can cognitively understand what the feedback signifies.

Warning strategies may utilize a sensing element. The sensing element detects a problematic condition and then either directly initiates a warning or informs a process controller that initiates a warning. The following example illustrates a mixed technique. It uses limit switches as a sensing element, an indicator light as a warning device, and does not allow the assembly process to proceed until it is done properly, that is, a form of forcing function.

The example is an assembly process and is taken from Shimbun [3]. The task has a worker assembling a kit. It is important that all the necessary parts are included in the kit. Before the intervention, a worker, due to fatigue or inattention, would omit parts or sometimes add extra parts. The before case had the worker reaching into inventory boxes and filling the kit. "Correct assembly relied entirely on worker vigilance" [3].

The process improvement used "smart" inventory containers in that the containers had a lid with a sensing switch and an indicator light. All the lights are initially off. When the worker opens the lid to a box, the switch senses this action and the container's indicator lights up. The workpiece does not advance until all the lights have been lit, that is, a proper assembly.

Assembly operations at all the major companies have similar *poka-yoke* devices. A variation of this kind of system is shown in Figure 10.4. This is a commercially available system using a "light curtain" sensing system. When the light beam is broken, it signals a worker's hand picking an object. If this is an incorrect action, an indicator light goes on along with a buzzer that warns the worker to correct the action.

Figure 10.4 A just-in-time inventory cart is instrumented with light curtain sensors. When a worker's hand enters the bin to get a part, it breaks the light beam. There is a light beneath each bin. If the assembly process uses barcodes or RF tags to identify an assembly, a processor controller can activate the light on the bins with the correct parts for that assembly.

The Enabling Technologies Laboratory (ETL) designed the task sequencing system (TSS) [12]. The TSS guides workers through assembly or packaging tasks. Using smart sensor devices, it supports auditory feedback as a prompting mechanism in addition to the feedback provided by light indicators. Workers with cognitive impairments often lose attention and require a prompt to refocus attention. The auditory prompt can be turned on or off as required. The ETL TSS is a *poka-yoke* device and exemplifies universal design in that it improves quality and performance for all workers. As a *poka-yoke* device, it fights boredom and fatigue in normal workers, but from a universal design perspective it provides feedback and guidance that allow cognitively impaired workers to do the job at competitive production rates.

Strategy 15.3: Provide quick and easy recovery if an error has occurred
If an error occurs, make it easy to undo the error, or correct a mistake once it has happened.

When we make an error using a computer word processing system, we typically have an option to undo a given action. For example, the user can *undo* a delete or copy operation. Windows operating systems use <Ctrl>z key sequence to undo keystrokes, which is another example. In this kind of process, the operation we performed was a legal operation, only the context or situation renders it an error. It is important that the system allows us to easily undo such errors.

For the counting and packaging systems just discussed, a warning signal indicated an error and if the person understood the feedback, a corrective action could easily be taken. Errors that result from slips—for example, inattention, fatigue, boredom or forgetfulness—can usually be corrected by

Chapter ten: Error-managed (proofed) 139

such warning strategies. The signal redirects our attention or helps us remember what to do. However, errors that occur from mistakes are more subtle and difficult to manage.

Mistakes

Errors that result from mistakes, that is, thoughtful, deliberate actions that are incorrect and lead to errors, are more difficult to address [10]. Mistakes can be reduced if the user has a good conceptual model of an entity's operations. Mistakes can be reduced if the information presented to a user is correct and communicated without ambiguity, vagueness, or errors. The user must be able to perceive the communications, have time to process and interpret the information contained within the communications, and have time to take appropriate actions. In short, mistakes can be reduced by the systematic application of the cognitive design strategies presented in the human factors chapter.

An example was given in Chapter 8, dealing with affordances and mappings, of a man who picked up a drink served in what appeared to be a heavy, thick-glass mug, but which was actually a light plastic mug. The result was that he lifted it like a heavy mug and got drenched. This is a classic example of a mistake. The affordance of the mug suggested one weight, and reality provided a message that the estimated weight was wrong, but alas, it was too late to correct the resulting error.

Using a word processor to create a document affords us the opportunity to make both slips and mistakes. Hitting the wrong key while typing a word is a slip. My current word processor can detect such spelling errors and, if the appropriate option is selected, automatically make the correction on-the-fly. If that is not possible, it will underscore the word with a red line indicating a possible spelling error. The potential spelling may or may not be an error. Spelling errors can be slips or mistakes, depending on the writer's state of mind. Additionally, the word processing system will underscore a word that is not in the dictionary; it may still be spelled correctly. If I know the word is correct, I can add it to my dictionary so that it is not flagged again as a potential error.

There are many cases where I want to write something that the system does not "like." For example, if, after hitting the <enter> key and advancing to a new paragraph, I want to start the new line with a lowercase letter, the system "corrects" the error on-the-fly by making the first letter uppercase. I know how to defeat this feature but sometimes such automatic "error-correcting" features are annoying.

Reading over a sentence or paragraph that I had written months ago and realizing that I had misinterpreted some data and drawn the wrong conclusions is an example of a mistake. At the time I wrote the paragraph, I deliberated over the words and the conclusion but I had misunderstood, misread, or been misdirected by the data. My current word processing system makes it easy to correct such mistakes. If I had used a manual typewriter to write the document, the correction of such a mistake within the context of a larger document would have been much more difficult.

The fundamental human factors principle for dealing with errors originating from *mistakes* is to build knowledge into the process. Examples of the application of this principle include the spelling auto-correction and underlying features that rely on a system dictionary. Grammar checkers rely on the standardized rules of grammar built into the word processing system as an "expert system." Another example of this strategy is word prediction associated with computer programs. As a person types, the program provides a dropdown box with a list of most likely words. For most people this is a nuisance unless it is part of the spellcheck feature. For people with certain cognitive impairments, such a feature allows them to use the software in a productive manner.

Palm-held devices often use a similar prediction strategy for calendar and contact information data entry. For example, when entering a new event into the calendar, Hewlett Packard's iPAQ provides a dropdown box with a list of past locations. Using the pointer and touch keyboard, the device uses word prediction and displays a high probability candidate word in a box at the bottom of the display. These features build knowledge into the device and help reduce errors.

Many software development programs have integrated dropdown help features into their products to help reduce programming errors associated with command, syntax, and formatting statements. One example is Microsoft Visual Studio. When the programmer enters the command "Dim test As Int", it says to dimension the variable "test" as an integer "Int." However, there are many types of integers and the system provides a dropdown box on the screen, next to the command, providing integer dimensioning options. As the programmer highlights each option (e.g., Int16), an explanation box is provided to the side. This element is part of the IntelliSense feature, which provides such information for system instructions and commands.

This example illustrates several universal design strategies. Microsoft Visual Studio is a very complicated and complex program to use [13]. The dropdown help or option boxes build knowledge into the system so that the user does not have to go into the Help System. The dropdown box provides the programmer with options for the specified command, thus reducing the probability of an error. If the dropdown boxes are insufficient, then the users can go to the Help System, which provides more detailed information.

Error-proofing for training and educational processes

The opening discussion on error-proofing noted that in an educational or training system, where people learn by making mistakes, one should error-proof those elements of the process not critical to the learning objectives, and then manage the errors in the process segments not directly

Chapter ten: Error-managed (proofed) 141

```
        ┌──────────────┐
        │  Setup &     │
        │ preparation  │
        └──────┬───────┘            Manage errors
               │                         ↙
    ↗          ↓
Error-proof  ┌──────────────┐
             │ Educational  │
             │   activity   │
             └──────┬───────┘
                    │
                    └──────→ ┌──────────────┐
                             │  End activity│
                             │  transition  │
                             │   clean up   │
                             └──────────────┘
                                   ↗
                               Error-proof
```

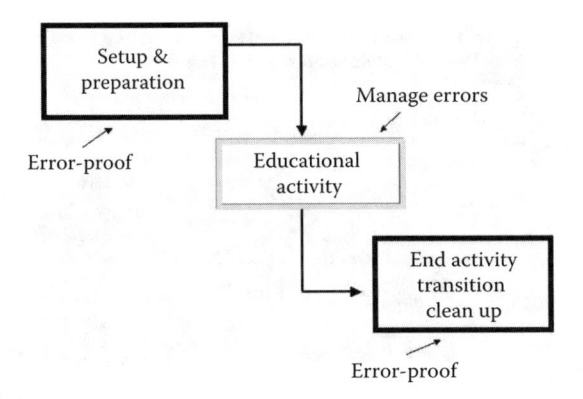

Figure 10.5 People learn by making and correcting mistakes. Manage errors for that part of the process. Error-proof the remaining steps to avoid non-value-added activity and wasted time and energy.

associated with the educational objectives so as to facilitate the educational process. Figures 10.5 through 10.8 illustrate this idea.

Figure 10.5 illustrates the basic idea. Manage errors during the educational activity and error-proof the other non-value-added activities, that is, activities not directly related to the educational activity. Figure 10.6 illustrates this idea for the educational activity of writing a report, using a computer, in a school classroom. Figure 10.6 shows the essential elements of a computer set up for a writing class. The value-added or educational activity is the *writing process*, the box with a heavy border. All the other activities are considered non-value-added with respect to the creative act of the *writing process*.

Computer access must be guaranteed accessible for all students. This access could include the standard keyboard/mouse arrangement as well as a variety of assistive technologies. All students need access to the application program, in this case a word processing program. This means selection of a word processing program and selection of all required assistive technology. Note that the assistive technology changes at each point in the process. The next stage is access to the specific word processing system. Again, this involves a variety of system configurations and integration with assistive technologies. If all these stages are not error-proofed, then the students and teachers may spend a lot of non-value-added time and energy finding hardware or software, problem-solving, and fixing problems. These non-value-added activities are associated with tasks not directly related to the creative act of writing. They lead to student and teacher frustrations and possibly never getting to the value-added task of writing.

When the student is finished writing, the resultant document needs to be delivered to the teacher for assessment and comments. As noted in Figure 10.6, there are many ways this can be done, and all necessary options should be error-proofed to, again, avoid the frustrations and issues related to the

142 Universal and Accessible Design for Products, Services, and Processes

Accessibility concerns associated with writing a report at school in a classroom

```
┌──────────┐     ┌───────────┐ ┌──────────┐     ┌──────────────┐
│ Computer │────▶│Application│ │ Working  │────▶│Access to program│───┐
│  access  │     │  program  │ │ document │     │functions and features│  │
└──────────┘     │   access  │ │  access  │     └──────────────┘    │
                 └───────────┘ └──────────┘                              │
```

Physical access:
>Keyboards/mouse
>Dictation
>Intellikeys
>Dynavox
>Switch/visual scanning
>Mouth stick
>Head pointer
>Eye gaze

Access to:
>Operating system
>Words/icons/color
>Simplified structure
  —Overlays
    +Intellitools
    +Dynavox
  —Directly to Word
    processor

>MS Word
>Appleworks
>Co-Writer
>Write out loud

Access to functions & features of the program.
>Built into the program
  How easy to set-up?
    +Font size
    +Color
    +Spelling
    +Grammer
    +Word prediction
    +Dictation
>External/added
    +IntelliTools
    +JAWS/screen reader

```
┌──────────────┐    ┌──────────┐┌──────────┐    ┌──────────────┐
│ Retrieving & │    │ Printing ││Access to │    │Writing process:│
│delivering the│◀───│   the    ││ printers │◀───│learning/creative│◀──┘
│  document    │    │ document ││          │    │     act      │
└──────────────┘    └──────────┘└──────────┘    └──────────────┘
```

>Physically get and deliver
>Email
>Fax

How is the print function envoked and executed?

Cable/infra red/wireless?

This is the value-added part of the process.

Figure 10.6 The educational focus of this activity is expository or creative writing, and that activity has a thicker outline. The remaining activities should be error-proofed so as to facilitate the education objectives.

non-value-added tasks of problem-solving and fixing printer problems, student accessibility to a printer, or mode of distribution.

Figure 10.7 shows the same diagram but with a different educational objective. Here, the value-added activity is to learn how to ensure accessibility and use of a specific word processing program. The associated task boxes are outlined with a heavy border. In this case, the learner is allowed to make mistakes in many more tasks. However, it is still important to error-proof the remaining non-value-added tasks.

Figure 10.8 shows the same diagram, only here the educational activity is to learn about the accessibility and use of computer-based word processing systems. The educational goals now encompass all the activities. The learner is now free to make errors in all the associated activities.

All training and educational processes should clearly identify their educational goals and objectives. This will define what elements of the overall system are value-added or non-value-added components. The non-value-added components must then be error-proofed and the core educational activity error-managed for effective learning to occur.

Chapter ten: Error-managed (proofed) 143

**Accessibility concerns associated with learning
how to use a specific word processing program**

| Computer access | → | Application program access | Working document access | → | Access to program functions and features | → |

Physical access:
>Keyboards/mouse
>Dictation
>Intellikeys
>Dynavox
>Switch/visual scanning
>Mouth stick
>Head pointer
>Eye gaze

Access to:
>Operating system
>Words/icons/color
>Simplified structure
  −Overlays
    +Intellitools
    +Dynavox
  −Directly to Word processor

>MS Word
>Appleworks
>Co-Writer
>Write out loud

Access to functions & features of the program.
>Built into the program
   How easy to set-up?
      +Font size
      +Color
      +Spelling
      +Grammer
      +Word prediction
      +Dictation
>External/added
      +IntelliTools
      +JAWS/screen reader

| Retrieving & delivering the document | ← | Printing the document | Access to printers | ← | Writing process: |

>Physically get and deliver
>Email
>Fax

How is the print function envoked and executed?

Cable/infra red/wireless?

This is now part of the process.

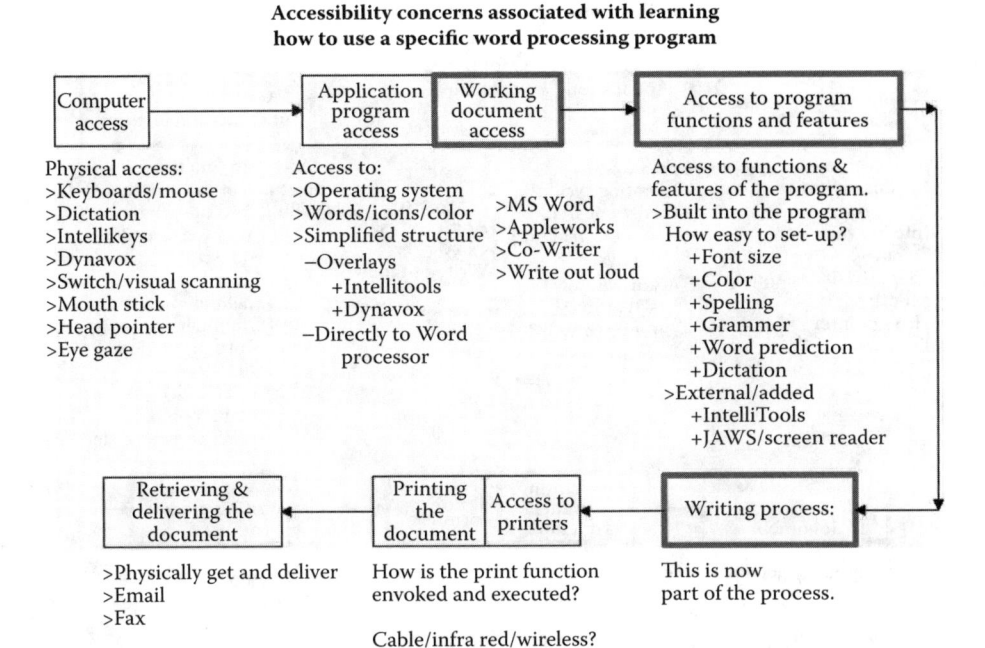

Figure 10.7 Here the educational focus is to learn how to ensure accessibility and use a specific word processing program. Because the learning objectives include more activities, there are more opportunities for errors. The associated activities have a thicker border. The non-value-added activities should be error-proofed.

Conclusions

People make errors. People are forgetful. Error-proofing strategies are employed to reduce or eliminate errors but the implementation of these strategies also yields processes that are more efficient, safe, and predictable. These are beneficial synergistic effects.

When someone makes an error, it must be corrected and/or the effects of the error must be fixed, taking time away from the desired task or objective, and introducing inefficiencies. Errors can lead to accidents where people can be hurt, and facilities and equipment destroyed. From a process perspective, accidents and their consequences create inefficiencies and unstable and unpredictable environments and processes. The occurrence of errors can render a process unpredictable in that one never knows when something will go wrong.

Error-proofing is a process-oriented universal design principle. The systematic utilization of error-proofing strategies leads to a variety of synergistic effects. These will be considered in chapters 11 and 12.

Accessibility concerns associated with learning about computer word processing

```
[Computer access] → [Application program access] [Working document access] → [Access to program functions and features]
```

Physical access:
>Keyboards/mouse
>Dictation
>Intellikeys
>Dynavox
>Switch/visual scanning
>Mouth stick
>Head pointer
>Eye gaze

Access to:
>Operating system
>Words/icons/color
>Simplified structure
  —Overlays
   +Intellitools
   +Dynavox
  —Directly to Word processor

>MS Word
>Appleworks
>Co-Writer
>Write out loud

Access to functions & features of the program.
>Built into the program
 How easy to set-up?
   +Font size
   +Color
   +Spelling
   +Grammer
   +Word prediction
   +Dictation
>External/added
   +IntelliTools
   +JAWS/screen reader

```
[Retrieving & delivering the document] ← [Printing the document] [Access to printers] ← [Writing process]
```

>Physically get and deliver
>Email
>Fax

How is the print function envoked and executed?

Cable/infra red/wireless?

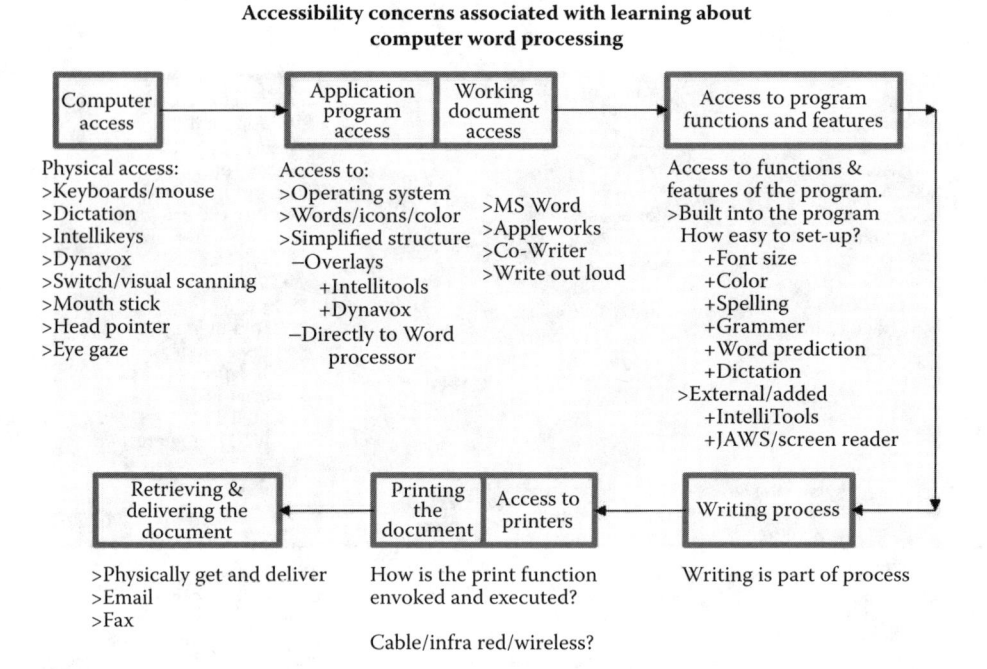

Writing is part of process

Figure 10.8 The educational focus in this diagram is even broader: accessibility to computer word processing systems.

References

1. M. Imai, *Gemba Kaizen*. New York: McGraw-Hill, 1997.
2. S. Shingo, *Zero Quality Control: Source Inspection and the Poka-yoke System*. Portland, OR: Productivity Press, 1986.
3. N. K. Shimbun, *Poka-yoke: Improving Product Quality by Preventing Defects*. Cambridge, MA: Productivity Press, 1988.
4. Access Board, *Telecommunications Act Accessibility Guidelines: Final Rule*. Washington, D.C.: *Federal Register,* 36 CFR Part 1193, 1998.
5. Access Board, *Electronic and Information Technology Accessibility Standards*. Washington, D.C.: *Federal Register,* 36 CFR Part 1194, 2000.
6. Japan Human Relations Association, *Kaizen Teian* 2, J. H. R. Association, Ed., first English ed. Portland: Productivity Press, 1992.
7. R. F. Erlandson and D. Sant, Poka-Yoke Process Controller Designed for Individuals with Cognitive Impairments, *Assistive Technol.*, 10, 102–112, 1998.
8. R. F. Erlandson, Universal Design in the Workplace, *The President's Committee on Employment of People with Disabilities*, Invited Panelist, 1998.
9. R. F. Erlandson, I. Kierstein, and D. McElhone, Assembly Techniques to Sheltered Workshops and Community Worksites: Working Smarter not Harder, presented at *Council on Exceptional Children – Annual State Conference*, Grand Rapids, MI, 1998.

10. D. A. Norman, *The Design of Everyday Things* (originally published in hardcover by Basic Books, 1988). Basic Books – Member of the Perseus Book Group, 2002.
11. howstuffworks.com, How does a gasoline pump at a filling station know when the tank is full?, howstuffworks.com, 2004.
12. ETL, Task Sequencing System, http://webpages.eng.wayne.edu/etl/Products/TaskSequencing.html, 2000.
13. Microsoft, Visual Studio 2005: provides a range of tools that offers many benefits for individual developers and software development teams, http://msdn.microsoft.com/vstudio/vshome.aspx, 2005.

chapter eleven

Efficient (muda elimination)

Chapter goals:

- State the universal design principle dealing with efficiency and provide design strategies.
- Define non-value-added activity in relation to efficiency.
- Provide example applications of the strategies.

Efficient (muda elimination)

Principle

Designed entities need to be efficient in that they have reduced as much of the non-value-added activities as possible and/or is reasonable. In *kaizen* terms, one would say *muda elimination* [1]. *Muda* means waste in Japanese; however, "the implications of the word include anything or any activity that does not add value" [1].

Discussion

Non-value-added activity (NVAA) is an important process concept. NVAA is any activity that does not directly add to the successful and timely completion of the task or activity [2]. Like the other universal design process principles, efficiency is a broad concept that places constraints on a number of human factors principles and strategies. One factor is the use of ergonomic principles that reduce the NVAAs of carrying, lifting, and material handling [2]. Efficiency includes the reduction of storage and waiting times [2]. It includes the efficient use of natural resources and consumable materials [2]. For manufacturing or assembly jobs, non-value-added activity has been classified according to seven categories: (1) overproduction, (2) inventory, (3) repair/rejects, (4) motion, (5) processing, (6) waiting, and (7) transport [1]. Similar non-value-added categories can be associated with service-sector jobs.

Most discussions of universal design do not include efficiency, or alternatively *muda* elimination. *Muda* elimination has typically been associated with *kaizen*, continuous improvement, techniques for the workplace, but *muda* elimination is a powerful and universal concept that touches on issues not raised by more traditional discussions of universal design.

From a building or facilities point of view, *muda* elimination would include providing doors, walkways, elevators, and escalators so as to allow an efficient flow of people for normal operations. It means placing handicapped parking reasonably close to curb cuts, ramps, and building entrances so that people using handicapped parking spaces do not have to negotiate long and circuitous routes to gain access to the building or facilities.

Consumers want products that are easy to use and do what they are supposed to do. If something keeps breaking down or malfunctions or is too complex or confusing to operate, if you constantly have to refer to the users manual or keep making "silly" mistakes, you soon stop using the product. Consumers do not want to engage in a lot of wasteful and frustrating NVAA.

Viewed as the elimination of anything or activity that does not add value, *muda* elimination has potential utility for consumer product design. For example, studies have shown that people use relatively few of the hundreds of options available with most commonly used computer word processing programs [3]. Does the added complexity associated with a proliferation of added features add value to a product? In his book entitled *Crossing the Chasm*, Moore presents a model of consumer buying behavior with respect to technology products [4]. The largest percentage of consumers are not interested in all the bells and whistles; they are looking for a product that will address their relatively narrow set of problems. The added complexity resulting from a large collection of mostly never-used features not only does not add value, but for many people actually reduces value [4].

Similar issues of choice arise with the choice of service options, such as which mobile carrier to use, which retirement plan, which medical plan. In the *Paradox of Choice*, Schwartz makes the argument that contrary to popular opinion we can indeed have too many choices [5]. Furthermore, the proliferation of choices can lead to a decision anxiety wherein people simply do not make a choice. Schwartz cites public opinion surveys that show that the "majority of people want more control over the details of their lives, but a majority of people also want to simplify their lives" [5]. This, according to Schwartz, is the paradox of our time [5]. This paradox presents an interesting challenge to designers.

Service operations should be designed to provide fast, yet error-free, service to clients. Waiting for a person on a phone-based ordering or information service is non-value-added activity for the client. Sitting in a doctor's office waiting to see the doctor is non-value-added time for the patient. It is common practice to design such cueing systems to maximize the use of the expensive human resource—that is, sales agent, information provider, or physician. Such a design places much less value on the consumer or client.

Breakdowns and poor maintenance operations lead to non-value-added activity and inefficiencies. For example, a person in a wheelchair who has to wait for four buses to pass by his stop before one comes along with a wheelchair lift that works exemplifies non-value-added activity for the consumer.

Work can be viewed as a series of processes or steps. Work steps can add value to the product. For example, in manufacturing, a piece of steel is milled into a shape. In an assembly task, parts are joined together. In the service sector, information is added to a document, French fries are added to the customer's plate, or purchased merchandise is placed into a bag for the customer. Other steps do not add value to the product. Such non-value-added steps include parts waiting for transport to the next manufacturing or assembly step, lifting heavy parts from the floor to a tabletop, a restaurant dinner order sitting on the counter getting cold while waiting for a server to take it to the customer, or inspection of a part after an assembly operation. Hence, process resources, which are people and/or equipment, either can add value or do not add value.

Design strategies

What makes something easy to use? Certainly ergonomic considerations such as control placement, knobs that turn easily, etc., are factors. People need good, reliable affordances, mappings, and feedback, that is, they need knowledge built into the system. People should be able to perceive feedback and be able to interact with the product.

Strategy 16: Reduce or eliminate non-value-added activity (NVAA)

This is the fundamental strategy with respect to efficiency. Non-value-added activity covers a broad spectrum of process functions. This strategy complements many of the other strategies, and process simplification is often a result of reducing or eliminating non-value-added activity because task or activity elimination is the main objective.

Strategy 16.1: Use quality control and reliability engineering techniques to reduce or eliminate NVAA due to repair and rejects, and excessive waiting due to setup and equipment breakdown

Quality control and reliability engineering techniques are designed to reduce or eliminate equipment breakdown and process failures. Their application is intended to yield products and processes that are reliable, and easy to set up and configure.

Error recovery and correction is another form of NVAA. The cognitive elements of design address non-value-added activity such as errors and the time and resources required to correct for or undo them; this could include re-booting or re-starting the system, having to make equipment repairs or replacements. It would include non-value-added activity such as frequently having to stop to read a help menu or the instruction manual because the affordances, mappings, constraints, and feedback do not provide sufficient knowledge and information for product or process operation.

> *Strategy 16.2: Make the designed entity as simple and easy to use as possible*

The efficiency principle and this design strategy integrate elements from all the previous principles and strategies, but do so with a broader product, process, and systems perspective. It is the integrated and cumulative effect of these human factors principles along with flexibility and error-proofing that culminate in an entity that is easy to use and simple. The design emphasis on easy-to-use and simple brings to the fore the balance a designer faces between designing a general-purpose device that offers wide functionality and options versus designing an appliance or a device targeted at one or two specific functions with few if any frills or options.

Appliances versus general-purpose devices

The efficiency principle and this strategy bring into a focus a central issue with respect to the design of electronic and information technology, that of appliance versus general-purpose device. The microprocessor, nano-devices, and the very large scale integration and packaging of components has and continues to revolutionize electronics and information technology industries. Designers can provide options, features, and functions in profusion for little additional cost. The choice from a design point of view is to design devices that address a very specific function and nothing else, or design a general-purpose device that embodies a variety of functional capabilities.

In his book entitled *The Invisible Computer* [3], Norman reproduces a page from the 1918 Sears Roebuck catalog advertising a general-purpose electric motor system. The core was an electric motor that ran off home electricity. The motor came with a variety of attachments that allowed one to configure a fan, a grinder, a churn, beater, a sewing machine, and more than 15 other applications (see Figure 3.1 in [3]). Today we smile at such an advertisement. Electric motors are embedded in thousands of products and devices, from fans to watches, door openers, car window openers, power tools, and on and on. The electric motor has become invisible as an embedded element in an appliance. This is what is happening with the application of microcomputers in E&IT (electronic and information technology) products. However, the design inclination to add features and options has remained strong.

As an example, consider the design challenges associated with the human interface for the variety of mobile electronic devices, cell phones, PDAs, GPS devices, watches with a multiplicity of functions, navigation devices, etc. Designers are "looking at" a shrinking interface display. Designing a robust intuitive user interface is one of the most difficult engineering design tasks [6]. All the perception issues emerge from being able to see the display (size, contrast, brightness in different environments), hearing auditory cues, prompts, or messages, and having the system understand voice commands, to the size of the keys and control buttons, and the size, placement, and density of the icons for stylus touch options. Complexity is another issue; how many and what options and features will be supported? The technology can support more options and features than anyone could ever use.

Norman argues that packing too many features and options into one device is inefficient for users in that they will never use all the features and the features build in too much operational complexity [3]. Norman's argument complements that of Moore's in *Crossing the Chasm* [4]. Norman argues from a usability point of view and Moore from a marketing perspective. To counter the proliferation of features, Norman argues for the design of appliances, that is, devices designed for one specific function that are essentially plug-and-play.

Recall the Freeplay radio (see Figure 2.2). It is easy to use and exemplifies efficiency and universal design in a number of ways. It can be considered an appliance. The original radio was to work without the need for main power or batteries. The latest radio models also have solar panels and rechargeable batteries [7].

The following is an advertisement quote from the Freeplay website. "Freeplay's newest radio is the ideal portable product that you can take with you anywhere. Designed for people who want to enjoy the outdoors and who appreciate the security of a failsafe radio, this radio is fun and convenient, anytime, anywhere" [8].

The radio is easy to use, having only the essential controls: power (on/off), volume, and tuning. It is efficient in its use of power and also flexible in terms of its source of power (wind-up, solar, and rechargeable batteries.). Because of its ease of use, durability, flexible power requirements, and reliability, the Freeplay radio has become very popular with hikers, campers, mountain climbers, and people in remote areas.

Another appliance is the LilTrak 255 pedometer with an embedded microprocessor. This device allows the user to calibrate one's stride length to inches, and the device counts steps and converts them to distance walked. The user can display steps or distance (in miles or kilometers). The device also contains a stopwatch. All the measured elements can be reset to zero. There are three control buttons. A three position slide switch is used to select stop-watch, number of steps, or distance. There are two large buttons, one on either side of the device. One always resets the display, regardless of the unit being displayed. The other button only works in the distance mode and pushing it enters the stride calibration procedure. This device could have been designed to be much more complicated and offer many more options and features. In its current form it is an appliance designed for three essential functions.

Process and service considerations

Process and service applications exhibit NVAAs in the form of waiting and excessive transportation or travel. Wheelchair users and those with mobility problems, those using crutches or walkers, often experience unnecessary NVAAs. If handicapped parking is not close to an accessible entrance, then the customer may have an unnecessarily long route from parking place to entrance. Often there is no curb cut in close proximity to the parking place and the customer must first find and then travel to the curb cut to gain access to the building. Wheelchair-accessible elevators may be inconveniently placed. Wheelchair-accessible toilets may be few and far between, thus causing the consumer to spend time and energy finding them.

Strategy 16.3: Avoid complexity in that it leads to NVAA

Designers should use an appropriate level of language in the users manual, operating instructions, technical references, and support material. User activity such as rereading sections striving for clarity, looking up material in additional resources, and seeking additional human support are all forms of NVAA.

Phone service systems exemplify many NVAAs. First, the user is typically required to enter a series of numbers in response to questions and then after negotiating the "touch-and-go" process must wait until service is provided. If the response time is too short or the sequence of options so long that people cannot respond fast enough or forget what number takes you where, then the process fosters frustration and errors.

Designers should keep instructions and the sequence of operations as simple as possible. Recall that people have problems with activities that require remembering a lot of detail and alphanumeric codes or data. For example, some older office phone systems used arbitrary numeric codes to perform call functions. With one such system, if while taking a call and another call arrived, the user received an auditory signal. To place the active call on hold and receive the incoming call, the user had to depress the "flash" button twice, and then enter the digits 115. The vender supplied a trifold, small-print, printed on both sides, folder, "quickie guide" to help the user. Needless to say, the system was a bust. People could not remember all the arbitrary codes and rather than engage in frustrating NVAA, did not use the features and switched to a new system as soon as feasible.

Word processing programs that provide spellcheck features improve user efficiency while reducing the probability of an error. Having to correct spelling and grammatical errors is an NVAA. Typically, the spellcheck feature has options that can be activated or deactivated by the user. One can select an automatic correction mode or use the spellcheck from the toolbar. The spellchecker also provides a rudimentary grammar check and recommended alternatives. For most people, such features are convenient but many people could not use the technology if these features were not present.

Chapter eleven: Efficient (muda elimination) 153

Having to use computer-based online help systems is an NVAA in that time spent looking through the help material is time not spent working on a task. The best approach is to design the entity so that the user does not need to consult the online help. If that is not possible, then build knowledge into the system to make the process simpler and quicker. One example of this approach is the Dynamic Help feature used in the Microsoft Visual Studio program [9]. When this feature is activated, the screen splits vertically, with the right-most window displaying the instruction manual. As the programmer enters commands and instructions, the Dynamic Help window changes the displayed topic to coincide with the current command or instruction. If the programmer needs more information, it can be presented in another window.

In the design of service operations, too many choices can be confusing to consumers; recall the "paradox of choice" [5]. While Schwartz presented the "paradox of choice," he also argued for the following strategy to deal with choice: "Choice within constraints. Freedom within limits" [5]. Service designers must avoid the temptation to offer the client everything. In *Paradox of Choice*, Schwartz presents an example wherein a mid-sized accounting firm increased the number of retirement plans from 14 to 156. The employees had a very difficult time making a decision. Many did not make a decision for a very long time. People spent a lot of time studying the plans, trying to figure out which was best for them. Up to a point, such study is valuable. However, if dragged out, such study is clearly an NVAA. Another way to describe the "Choice within constraints. Freedom within limits" design strategy is to design a decision tree for service options that is not excessively deep or wide.

Strategy 16.4: Use task analysis techniques to identify tasks or activities that can be eliminated or redesigned so as to reduce or eliminate NVAA

Task analysis tools are designed to identify tasks that exhibit unnecessary motions or energy, tasks that serve no real purpose, tasks that are too complicated, tasks that exhibit excessive waiting due to setup and equipment breakdown, and/or tasks that exhibit transport problems such as poor timing, too frequent or infrequent.

For example, an assembly operation has the worker walk to an inventory storage area behind his workbench and pick up a large, but lightweight, plastic frame and place it on his workbench. The worker then reaches for other parts stored in bins within easy reach. All these task elements have been designed so as to be ergonomically sound. However, in this process the worker having to walk to get the plastic frame and pick it up and carry it to his workbench is non-value-added activity.

An agile system design might have the large plastic frames stored in a device that is near the workbench. The device not only stores frames, but also has a mechanism whereby the worker can push a button and the frame is dispensed into position on the workbench. Such a design is more efficient in that the non-value-added activities of walking, picking, carrying, and placement have been eliminated or greatly simplified.

The job is also more accessible because the physical demands of the job are greatly reduced. With the new agile device, perhaps a person in a wheelchair can now do the job. Such job redefinitions are critical with respect to Title I of the ADA, wherein the concept of a job's essential functions is defined. In the original process, it would be natural to say that being able to walk, pick, and carry the plastic frame are essential functions of the job. However, with the redesigned job, those are no longer essential functions of the job.

Inspection of process operations or products is a non-value-added activity [2]. The counting example (Figure 10.2) and the kit assembly job (from Chapter 10) illustrate jobs where the "before improvement" process promoted miscounting errors. The "before improvement" processes required an inspection step at a subsequent station [10]. Another worker had to inspect the packaging or kit to make sure all the components were present and in the correct amount. After the improvement, the processes were error-proofed to the point that the inspection step was not necessary. Removal of the inspection step renders the overall job less complicated because there are fewer steps. The worker assigned to the inspection task can be reassigned to a more productive, value-added job.

Conclusions

These examples illustrate several points with respect to efficiency as a universal design principle. First, it focuses on the process. While individual task elements might be ergonomically sound, perceptible, and cognitively sound, they might not add value to the process and in that sense the entire task is a candidate for removal or redesign. Second, removing a process activity and/or redesigning the process typically makes the process simpler (fewer steps) and often reduces both the physical and cognitive demands of the process rendering the process more accessible. Third, the process redesign can redefine the essential functions of the job. Changing the essential functions of the job has legal implications with respect to accessible design as specified in Title I of the ADA.

It is virtually impossible to remove all non-value-added activity from processes or products. Some systems are just very complex and will require extensive help features and users manuals. It may be impossible to remove all carrying or lifting. It may not be practical to use forcing functions as an error-proofing strategy but rather provide warnings that an operator can choose to ignore. The goal of this universal design principle is to create the most efficient system, product, process possible, that is, feasible given all the design constraints.

References

1. M. Imai, *Gemba Kaizen*. New York: McGraw-Hill, 1997.
2. M. Imai, *Kaizen*, 1st ed. New York: McGraw-Hill, 1986.
3. D. A. Norman, *The Invisible Computer*. Cambridge, MA: MIT Press, 1999.
4. G. A. Moore, *Crossing the Chasm*. New York: Harper Business, 1991.
5. B. Schwartz, *The Paradox of Choice; Why More is Less*. New York: Harper Collins Publishers, 2004.
6. N. Cravotta, The shrinking interface paradox, *EDN: Techtrends*, July 24, 2003.
7. All That's Green, The Freeplay Story, http://www.allthatsgreen.ie/page8.html, 2002.
8. Freeplay, Freeplay advertisement, http://www.freeplay.net/newsite/product/product.html, 2004.
9. Microsoft, Visual Studio 2005: provides a range of tools that offers many benefits for individual developers and software development teams, http://msdn.microsoft.com/vstudio/vshome.aspx, 2005.
10. N. K. Shimbun, *Poka-yoke: Improving Product Quality by Preventing Defects*, Cambridge, MA: Productivity Press, 1988.

chapter twelve

Stable and predictable

Chapter goals:

- State the stable and predictable universal design principle.
- Define common cause variability and discuss its importance in terms of universal design.
- Present design strategies that work to reduce common cause variability.
- Provide examples of applying these design strategies.
- Demonstrate the hierarchical and synergistic relationships among the universal design principles and strategies.

Stable and predictable

Principle

Design entities to reduce common cause variation. That is, design entities to be stable and predictable so that users can expect performance that supports the desired activity.

Discussion

While each entity presents unique requirements for stability and predictability, a common theme across all designed entities is the need to reduce the inherent variability of using the entity.

Reducing the inherent variability in a process is a fundamental universal design principle and key to creating stable, predictable systems. Every task and process naturally has a certain amount of variability or variance. Deming termed this variability *common cause* [1]. Everyday occurrences, such as traffic volume and the timing of traffic signals, contribute to the time variability associated with one's drive to work. An accident or severe storm is an exceptional event that is not common to the process. Deming termed such events *special causes* [1]. In terms of design activities, designers need to plan for the occurrence of special cause events (malfunctions, breakdowns, component failures, fires, tornadoes), but they typically have no control over these special cause events. On the other hand, designers do have the ability to reduce common cause events and problems associated with common cause variability.

The task of balancing a broom on the open palm of one's hand, broom bristles up, provides an example of common cause variability. The laws of physics introduce a great deal of variability and can make this task quite difficult. An accomplished balancer has the expertise and skills necessary to balance the broom; however, most people would scurry around trying, to no avail, to keep the broom balanced. Most individuals simply do not possess the skills necessary to overcome the variation inherent in the process to successfully perform this job.

However, if the process is redefined by holding the broom in two hands and introducing some technology, such as bracing the broom handle against a tabletop, many more people can successfully perform the task. The redefined process—that is, the use of two hands to hold the broom and the use of technology (desktop)—enables improved performance and makes the task more accessible by reducing the variation inherent in the process (Figure 12.1). This illustrates a typical strategy for reducing the variability associated with processes: redefine the processes and incorporate enabling technology.

A stable and predictable designed entity takes on very different forms, depending on the context. For example, from a buildings and facilities perspective, control of the physical environment in terms of temperature, humidity, light, noise level, etc., is fundamental. There should be no wild temperature fluctuations; nor should there be severe fluctuations in light or noise level, or in electrical power. Elevators, escalators, and all infrastructure

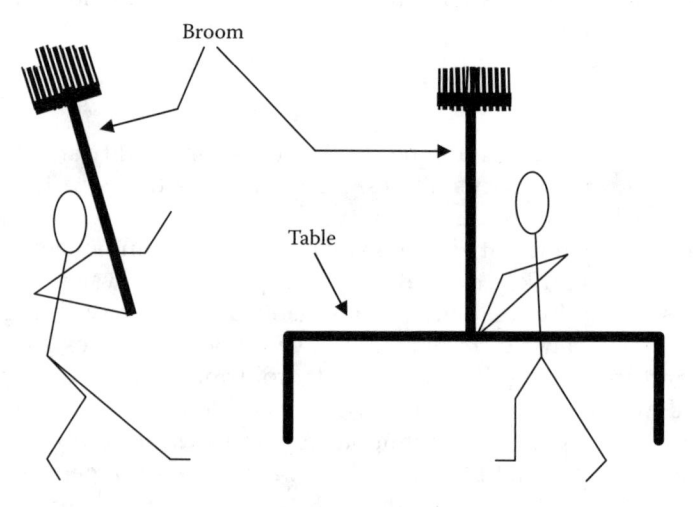

Figure 12.1 Common cause variability associated with balancing a broom versus the use of a table to help stabilize the broom.

systems should be reliable. The common cause variability of these elements should be very low, essentially zero.

From a driver's perspective, reduced inherent variability of an automobile means that first and foremost the automobile should be safe and reliable. It should not break down frequently; it should start every time; and the windows, radio, and other component systems should function properly. When considering a computer system, consumers also want it to function without software crashes or hardware malfunctions. The system should behave consistently. An application program should not introduce random events such as an unpredictable movement of the cursor or skipping lines. Users want stability and predictability.

Design strategies

Designers need to create entities that exhibit the least common cause variability as is reasonably possible. Common cause variability can be addressed at two levels: (1) macro and (2) micro. The macro level covers broad national and international standards, while the micro level targets individual products and processes. Figure 12.2 shows the strategies associated with this design principle. The macro level is discussed first, followed by the micro-level strategies. At the macro (national and international) level, the most powerful force for the reduction in variability is the creation of product, process and service standards. At the micro level common cause variability arises from three major sources: (1) the human-product/process interface, (2) the product, and (3) the process. While related, each of these sources of common cause variability has unique factors that warrant individual consideration. The discussion starts at the macro level.

Strategy 17: Work to establish national and international standards for products, processes, and services so as to reduce their common cause variability

From a national and global perspective, standards are essential for product development and sales—both nationally and internationally. In the early 1880s when light bulbs were first invented, bulbs and sockets came in more than 175 different sizes. Things were so confusing (highly variable) that by 1884 a standard had emerged for bulbs and sockets [2]. Another major problem in the late 1880s was the variability and lack of standards for threads on hoses, screws, and other industrial parts that were supposed to fit into one another [2]. It was not until 1901 when the U.S. Congress created the National Bureau of Standards (now known as the National Institute of Standards and Technology, NIST) in the Department of Commerce [3] that the variability and chaos started to be eliminated.

NIST is now an agency of the U.S. Commerce Department's Technology Administration. NIST was the federal government's first physical science

Principle design entities to reduce common cause variation.

Macro

Strategy 17 Work to establish national and international standards for products, processes, and services so as to reduce their common cause variability.

Micro

Interface

Strategy 18 Reduce the common cause variability associated with the person's interaction with the product or process.

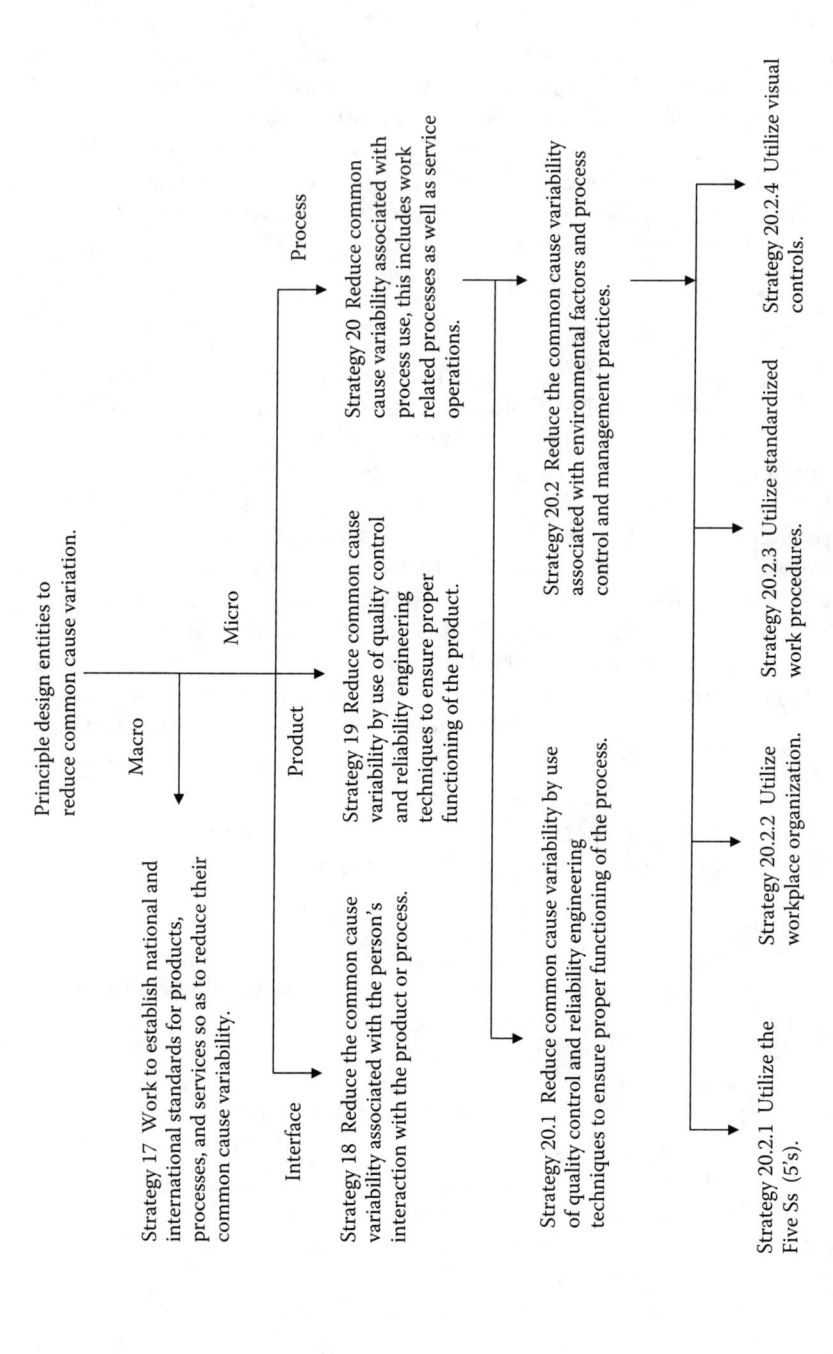

Product

Strategy 19 Reduce common cause variability by use of quality control and reliability engineering techniques to ensure proper functioning of the product.

Strategy 20.1 Reduce common cause variability by use of quality control and reliability engineering techniques to ensure proper functioning of the process.

Process

Strategy 20 Reduce common cause variability associated with process use, this includes work related processes as well as service operations.

Strategy 20.2 Reduce the common cause variability associated with environmental factors and process control and management practices.

Strategy 20.2.1 Utilize the Five Ss (5's).

Strategy 20.2.2 Utilize workplace organization.

Strategy 20.2.3 Utilize standardized work procedures.

Strategy 20.2.4 Utilize visual controls.

Figure 12.2 Common cause variability of products and processes can be reduced by a variety of strategies. Shown here are a macro-level strategy and a number of micro-level strategies. The micro-level strategies target the human–product/process interface, the product, and the process.

research laboratory. In the early 1900s, the United States had few, if any, authoritative national standards for any quantities or products. Regional and local standards were often arbitrary and conflicting. This confusion made it very difficult for American companies to get parts that fit together properly [3]. NIST worked to develop national standards that facilitated economic growth by reducing or removing the variability associated with no standards.

In 1918, the American Engineering Standards Committee (AESC) was formed to serve as the national coordinator in the standards development process as well as an impartial organization to approve national consensus standards and halt user confusion on acceptability [4]. The American National Standards Institute (ANSI) adopted its present name in 1969. ANSI is a private, non-profit organization that administers and coordinates the U.S. voluntary standardization and conformity assessment system [4]. ANSI standards cover virtually every product and service. ANSI represents the United States in the International Organization for Standardization (ISO), a worldwide federation of national standards bodies representing more than 145 countries. The ISO is a nongovernmental agency based in Geneva, Switzerland, with the mission of promoting the development of international standardization with a view to facilitating an international exchange of goods and services [5]. Without the standards developed by these organizations, there would be chaos in terms of product design and marketing. There would be no guarantee that a CD (compact disc) made in the United States would work in France.

In addition to product standards, there are also manufacturing and assembly process standards . Along with ANSI and the ISO, the IEEE (Institute for Electrical and Electronic Engineers) and the International Council on Systems Engineering (INCOSE) are very active in national and international process standards activities [6, 7].

For the remaining discussion of variability, it is assumed that national and international standards for products and processes are specified, and that as designers we must work within the constraints afforded by these various standards. Hence our attention will move to the micro level, that is, the design of specific products or processes.

Recall that a process can be defined as a collection of related tasks or activities that lead to results. Process tasks and activities can involve the use of devices. In this context, common cause variability can originate from the person, the product (device), or the process. Hence we need to look at design strategies that focus on this broad spectrum of causes—strategies that focus on the person/system interface, the product, and processes and services.

Strategy 18: Reduce the common cause variability associated with the person's interaction with the product or process

As designers we must deal with human variability; anthropometric differences, cognitive differences, and varying memory and motor skills, but we can influence the interface between the human and the product or process.

It is virtually impossible to remove the common cause variability associated with human performance. Each person will induce a variability that is associated with his or her own unique native endowments, skills, abilities, training, etc. However, a systematic application of the earlier design strategies will keep this variability to a minimum for each person.

A systematic application of the human factors design strategies constrained by the design's functionally demanded process requirements goes a long way toward reducing common cause variability associated with the human/system interface. In essence, employ appropriate human factors considerations, build knowledge into the product and process to guide the user in its operation, and use error-proofing techniques to reduce or eliminate user errors. Reduce complexity by identifying and eliminating non-value-added-activity (NVAA).

The common cause variability inherent in products (devices) and processes is considered next. The most fundamental strategy regarding the reduction of common cause variability for any product is to make sure that it is reliable and functions properly.

Strategy 19: Reduce common cause variability using quality control and reliability engineering techniques to ensure proper functioning of the product

This universal design strategy covers the more traditional concepts of quality control and reliability at both the macro and micro levels, that is, international and national quality control and reliability standards [8, 9], as well as high corporate standards and requirements [10–12]. Users expect products or process to do what they were designed to do, when they are supposed to do it, and do it without unexpected results.

Stability and predictability result from high-quality products and processes that are not only reliable, but also safe. Designers must be aware of a variety of organizations whose mission is to ensure consumer safety and product compliance with established safety regulations and guidelines. For example, the U.S. Food and Drug Administration (FDA) has announced its plan for strengthening the way it monitors the safety of medical devices after they reach the marketplace [13]. The Underwriters Laboratory, Inc. (UL) started in 1894 as the "Electrical Bureau of the National Board of Fire Underwriters" has evolved into an international testing and standards development laboratory for a wide range of consumer products [14]. Working in conjunction with ANSI, the ISO, and other standards-setting organizations, UL has developed more than 800 standards for product safety [14]. The UL website provides detailed information as to how a company may proceed to apply for and secure UL certification for its products [14].

Process tasks and activities can involve the use of products, that is, devices. A product can fail because a component breaks, a switch breaks, a wire comes loose, the soldering is bad, the fittings come loose, etc. A process can fail because one or more of the products or devices that are used in the process fail. Such reliability and process failure analysis flows from traditional engineering disciplines associated with complex system design and failure analysis. For example, the IEEE Reliability Society, with chapters in 60 countries worldwide, "is concerned with the problems involved in attaining reliability, maintaining it through the life of the system or device, and measuring it. The Society is engaged in the engineering disciplines of hardware, software, and human factors. Principal focus is on aerospace, communications, components and hybrids, computers, industrial electronics, lasers and electro-optics, medical electronics, nuclear and fossil power systems, and transportation systems. Reliability is integral to Availability, Maintainability, Quality, and System Safety" [15].

Processes also fail because of environmental factors, and process control and management practices. Strategy 20 covers processes, in general. The substrategies consider the different sources of common cause variability.

Strategy 20: Reduce common cause variability associated with process use, this includes work-related processes as well as service operations

It is axiomatic in process engineering and management circles that a system can neither be improved nor can a system's goals be achieved if the system is unstable, that is, exhibits too much variability [1, 16]. This strategy draws on traditional systems engineering strategies as well as all the preceding human factors principles and strategies and all the preceding process principles and strategies. This strategy emphasizes the synergistic nature of universal design. The systematic utilization of these previous strategies works to reduce common cause variability.

The first substrategy focuses on the traditional systems and industrial process engineering disciplines and methods.

Strategy 20.1: Reduce common cause variability using quality control and reliability engineering techniques to ensure proper functioning of the process

Systems engineering and industrial and manufacturing research are the traditional disciplines for the reliability, quality control, and analysis techniques referred to by this strategy [17–19].

The classic approach to the management of process variation is Six Sigma [20, 21]. Six Sigma and its variants form a vast body of knowledge, and therefore only a brief overview of the approach is provided. Six Sigma originally began as a method to reduce defects in production systems. Starting around 1986, Bill Smith at Motorola pioneered the use of metrics, developed using statistical techniques, for reducing production variability and hence limiting defects [21]. (Note that "Six Sigma" is a registered service mark and trademark of Motorola, Inc. [21].)

While Six Sigma had its origins in production defect control it has grown to include applications of variation management to all kinds of processes [22]. For example, healthcare [23, 24], banking [25, 26], construction [27], insurance [23, 28, 29], the military [30], and education [31–34]. Given this broad spectrum of application areas, two generalized approaches have emerged, one for the analysis and redesign of an existing process and one for the design of a new process. In the analysis and redesign of an existing process the DMAIC approach can be utilized; define, measure, analyze, improve, and control [21]. For new process design a variation called DMADV is used; define, measure, analyze, design, and verify [21].

For both approaches, the starting point is to carefully define the goals of either the improvement or the new design. In both approaches, the second step is to measure either the current state of affairs (a baseline condition against which to assess improvements) or for a new design to identify and estimate key process parameters. The third step in both approaches is analysis. For process improvement and redesign, it is necessary to verify relationships among the key process parameters. For process design it is important to develop and analyze design alternatives with respect to the key process parameters. This typically involves modeling and computer simulations. The approaches diverge at the fifth step: implement the process improvement for DMAIC and develop the final design for the DMADV approach. The final step for the DMAIC process improvement approach is to implement control and management procedures to ensure continuous improvement. The final step of the design approach is to verify that the new design is, in fact, doing what it was intended to do. This also includes the implementation of control and management procedures to ensure continuous improvement.

Both the DMAIC and DMADV approaches lay out very general steps. Note their similarity to the definition of design as specified by Miller [35]—"the thought process comprising the creation of an entity" presented in Chapter 2. DMAIC, DMADV, and similar design frameworks provide specific steps and procedures to Miller's very general definition. Such frameworks are necessary when dealing with complex design entities. In addition to design frameworks, designers need tools to help them deal with system and design complexity.

The tools and techniques utilized by the designers or planners vary, depending on the process and the step [1, 16, 36–40]. Automated production processes utilizing robotics and computer-controlled devices lend themselves to more mathematical and statistically sophisticated techniques, reminiscent of the original applications at Motorola. If, however, the process is an airline ticketing system or a telephone customer support service, the tools used reflect the more subjective nature of the process.

Probability theory is a fundamental tool in the analysis steps of both the DMAIC and DMADV approaches. A very simple example is presented to illustrate design insights provided by even elementary probability theory.

Chapter twelve: Stable and predictable 165

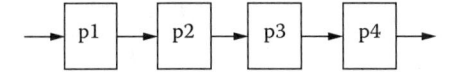

→ p1 → p2 → p3 → p4 →

Figure 12.3 A four-step process is shown. The probability of success for each step is shown: p1, p2, p3, and p4. Probability theory allows one to model the reliability of the overall process in terms of each step. The overall process is successful only if all the steps are successful.

Figure 12.3 illustrates a process with 4 sequential steps. Each step must be successful for the overall process to be successful, and each step's success or failure must be independent of the other steps. Let p1, p2, p3, and p4 denote the probability of success each of the steps. For simplicity, let all the probabilities of success be 90%, i.e., p1 = p2 = p3 = p4 = 0.9. In process reliability terms the reliability of each step is a relatively high 90%.

Based on this model, a designer can use probability theory to find the reliability for the entire process, that is, the successful sequential operations of all four steps. From probability theory, overall process success occurs when all the steps are successful. The probability of this happening is the product of all the step probabilities. Denote the reliability of the four-step process by R4; then R4 = p1 × p2 × p3 × p4 = 0.6561, or about 66 percent. Although the reliability of each step is high, 90 percent, the overall process reliability is low, 66 percent.

Figure 12.4 shows four independent processes in parallel. These processes all perform the same task, and each process has the same probability of success as before, 0.90. In this case, there will be a successful outcome if just one of the processes works. The overall process fails only if all the steps fail. The probability of failure for each step is 1 minus the probability of success. Denote the probability of failure for each step as follows:

q1 = (1 − p1) = 0.1, q2 = (1 − p2) = 0.1, q3 = (1 − p3) = 0.1, q4 = (1 − p4) = 0.1

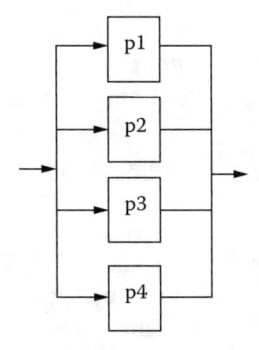

Figure 12.4 This figure shows a process with four parallel steps. The process is successful if at least one of the parallel steps is successful.

The probability of the overall process failing is the product of $q1 \times q2 \times q3 \times q4 = 0.0001$. The probability of overall process success, the reliability, is $1-.0001 = 0.9999$, or 99.99 percent.

The overall system has a higher reliability than any one of the individual subprocesses.

There are two obvious lessons a designer can learn from even such simple mathematical modeling. First, it is not wise to design a process with a large number of sequential tasks, even if each task has a very high probability of success. Second, redundancy, in the form of identical parallel operations, improves the reliability of a process. These are very general process design results obtained from basic probability theory. Process designers have much more powerful analysis and design tools at their disposal.

One example of such powerful design and analysis tools is the software provided by Dassault Systemes [41]. Dassault supplies a collection of computer-aided design (CAD) and analysis tools built around three-dimensional viewing of the products, processes, and process resources [41]. These software packages enable designers to design very complex processes and then view in three-dimensional animation the workings of that process, which can include human mannequins.

Design software such as offered by Dassault Systemes is a powerful enabling technology. The design software is enabling humans to deal with a level of complexity unheard of just 10 years ago with fewer errors and higher quality. Just as Six Sigma techniques have outgrown their manufacturing origins, so has traditional CAD outgrown its manufacturing origins. Virtually all traditional manufacturing companies in the automotive, airplane, shipbuilding, and railroad industries use such sophisticated CAD software. However, this design software has found applications in clothing and furniture design, jewelry design, architecture, theater stage and production design, and choreography [41, 42].

The next strategy focuses on environmental factors and process control and management practices.

Strategy 20.2: Reduce the common cause variability associated with environmental factors and process control and management practices

As noted previously, universal design is an umbrella term that covers a variety of established disciplines and expertise. Lean production, total quality management (TQM), and the granddaddy of them all, the Toyota production system (TPS), are such methods [11, 36, 43–45]. TPS, TQM, and lean production methods deal with environmental factors and process control and management practices. These techniques lead to improved product and process quality, a reduction in non-value-added activities, and error reduction. Most importantly, however, is the synergistic power of these seemingly simple techniques. This synergism can be better appreciated now that we have discussed the essential universal design principles.

To illustrate the synergism, consider two universal design principles that coincide with lean production and TQM. One of the fundamental principles is muda (waste) elimination [36, 37]. This is essentially the efficiency principle of Chapter 11. The key idea is to reduce or eliminate non-value-added-activity (waste). Error-proofing is another principle common to lean production/TQM and universal design. By reducing errors, the designer reduces the non-value-added-activity (NVAA) associated with correcting the consequences of making errors. As noted previously, removing or reducing the NVAA ergonomic activities, such as unnecessary lifting, carrying, or moving, highlights the hierarchical nature of universal design principles. While the ergonomic design of a lifting task or a carrying task might be within all acceptable OSHA requirements with respect to safety, the functional requirements of the task might be achieved without the worker having to lift or carry. From the efficiency principle perspective, lifting and carrying are NVAAs and subject to removal; that is, process principles place constraints on the lower-level human factors principles. Eliminating errors and NVAAs tends to reduce complexity and simplify the process, which in turn tends to reduce common cause variability.

This discussion considers four lean production/TQM strategies: (1) 5S, (2) workplace organization, (3) standardized work, and (4) visual controls. These strategies are process control or process management techniques. While focused on business-oriented objectives such as improved quality, improved production, and reduced costs, which are achieved in large part by a reduction in common cause variability [1, 36, 37], these same methods also reduce the physical and cognitive demands of a process, thus rendering them more accessible for a broader spectrum of people [46, 47].

Strategy 20.2.1: Utilize the five Ss

The Five Ss basically state good housekeeping techniques. The 5's are five action verbs all starting with an S in Japanese (Seiri, Seiton, Seiso, Seiketsu, Shitsuke) [37]. In English the five actions are Sort, Clean, Set in order, Standardize, and Progress.

Figure 12.5 suggests that the application of the 5's lays the synergistic groundwork for the application of other universal design strategies. If an environment is clean and organized, then one does not have to lift and move items in search objects. A clean, organized environment allows a designer to build knowledge into the environment without being obscured by clutter or misplacement.

Efficiency, ergonomics, and building knowledge into the process or environment by using error-proofing strategies have all been discussed. The application of each of these techniques individually brings benefits, but together there is synergy. Strategies for building knowledge into the environment and processes have been presented, from the human factors perspective in Chapter 8 and process perspective when dealing with efficiency and error-proofing. Three more strategies are now presented which, when implemented, build knowledge into the process or environment.

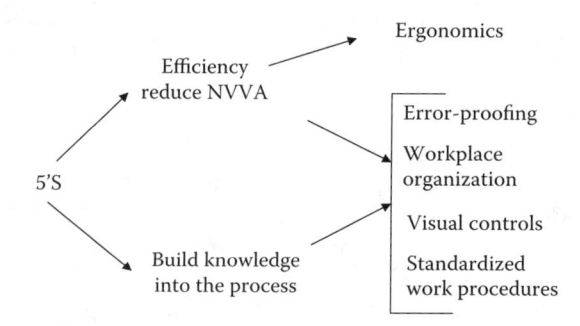

Figure 12.5 The 5S strategy deals with sorting, cleaning, setting things in order, standardizing, and continuous improvement or progress. Implementing a 5S program interacts synergistically with other universal design strategies.

Strategy 20.2.2: Utilize workplace organization

Workplace organization is another way to build knowledge into the environment and create more self-managing processes, thereby reducing common cause variability.

Designers need to utilize design principles that reduce the common cause variability inherent in work environments. Providing a safe, clean, and comfortable facility is an obvious and critical step [37]. Providing environments that are more self-managing is another step. Figure 12.6 shows a shadow diagram pegboard for tool storage. This is a simple self-organization

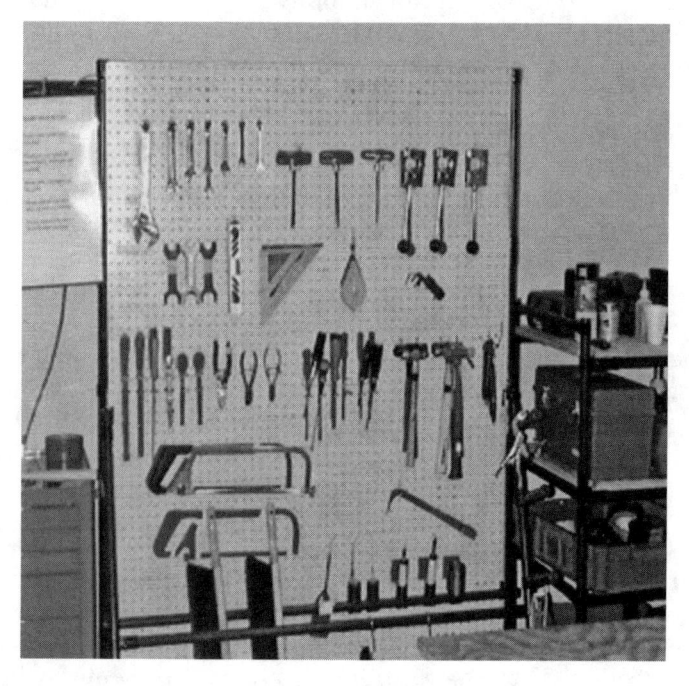

Figure 12.6 Storage unit using a shadow diagram for tool location.

mechanism. This simple technology provides significant self-managing features as well as error-proofing. Most people know by looking at the pegboard if a tool is present. Most people know to put the tools back over the corresponding shadow diagram.

Chefs use the term mise en place. This is a French term meaning to have all the ingredients necessary for a dish prepared and ready to combine up to the point of cooking—that is, workplace organization.

Strategy 20.2.3: Utilize standardized work procedures

At the corporate job level, standardized work procedures are designed to represent the best, easiest, and safest way to do a job; provide a basis for training; and provide a means for preventing the recurrence of errors and minimizing variability [36, 37].

In the martial arts, a kata is a sequence of prearranged movements or techniques [48]. Katas are a training tool; they embody the essential blocks, kicks, and movements associated with the specific discipline. Katas are classic examples of standardized work procedures. By mastering the kata, a practitioner masters the fundamental elements of the martial art.

Standardized operating procedures are similar to katas in that they are meant to coordinate the work effort and bring about a situation in which workers automatically, but based on an accepted rationale factory operation is inherently flexible and adaptable [16, 49].

For nonimpaired workers, standardized work should result in efficient, safe, cost-effective job processes. For the cognitively impaired worker, standardized work procedures create a stable, predictable work environment wherein the worker knows exactly what should be done and when it should be done. For many cognitively impaired workers, this structure allows them to succeed and perform competently at a job [47]. For all workers, standardized operating procedures reduce the common cause variability associated with the work.

Strategy 20.2.4: Utilize visual controls

Visual controls are used to build knowledge into the environment. As such, they serve to reduce error, improve efficiency, and reduce common cause variability [37].

Visual controls employ the visual perceptual and representational strategies presented in the human factors section, but visual controls convey information. Visual controls must communicate information to a person to be effective. Hence, the designer must make sure that the visual materials can be perceived. The visual materials should clearly represent unambiguous information. The function, purpose, and goals of a system inform the designer as to what is important. The human factors capabilities of the human agent inform the designer as to the nature of the visual materials necessary to convey the desired information.

The shadow diagram pegboard in Figure 12.6 is a visual control that organizes the environment. To ensure that the shadow diagram pegboard approach is successful, the workers must understand the meaning of the system, that is, the tools go back to the pegboard and in the spot indicated by the shadow diagram of the tool.

Parking lots offer another set of opposing examples. Two different examples are presented. At many county fairs, farm auctions, or community sporting events, people park their cars in an empty field. People try and line up their cars in rows but usually the rows are wavy and disorganized. Parking in a field without supervision illustrates a parking process with the largest, of the two examples, common cause variability. In shopping malls and parking structures, there are also usually no attendants, but there are many visual controls that organize and render the parking process in the parking environment self-managing with relatively little common cause variability. The parking spots are clearly marked by painted lines. There are directional arrows on the floor, walls, and often hanging from the ceilings. The floors are often color-coded and the sections numbered. Frequently, standard traffic signage also is present: stop signs, yield signs, no-parking signs, etc. Modern parking structures are marvels of self-managing processes using workplace organization, standardized work, and visual controls.

The parking structure works because drivers understand what all the representations mean, what they communicate. Traffic signs and rules are classic examples of standardized operating procedures. They reduce traffic common cause variability. Think of driving without such standardization.

Examples of the strategies and synergies

This section presents two examples that illustrate application of the principle to reduce common cause variability. The first is a meter reading job, and the second is an airport terminal.

Example 1: Meter reading

This example demonstrates a variety of universal design principles and as such emphasizes the synergistic effects possible. Suppose the job is to ensure that a device is operating within specified limits as indicated by meter readings. Figures 12.7(a) through (c) illustrate three different strategies termed Level 1, Level 3, and Level 4. The strategy used in Level 1, Figure 12.7(a), is to have a person read the meters and record the values on a sheet of paper. Then the person goes back to a reference book and checks to make sure the recorded values are within the specified ranges. This is an error-prone process. The person must be able to read, write, and interpret or approximate the value of each meter. Judgments such as interpretation and approximation are highly error-prone processes with a large amount of inherent variability. The process is ergonomically bad in that the person must

Chapter twelve: *Stable and predictable* 171

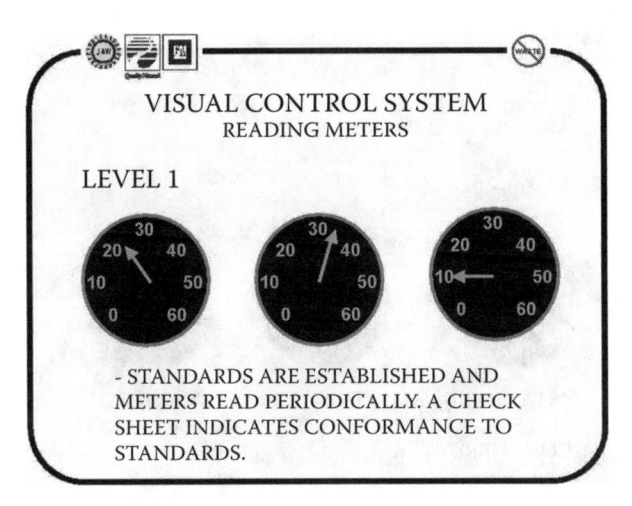

Figure 12.7 (a) This scheme has the highest common cause variability.

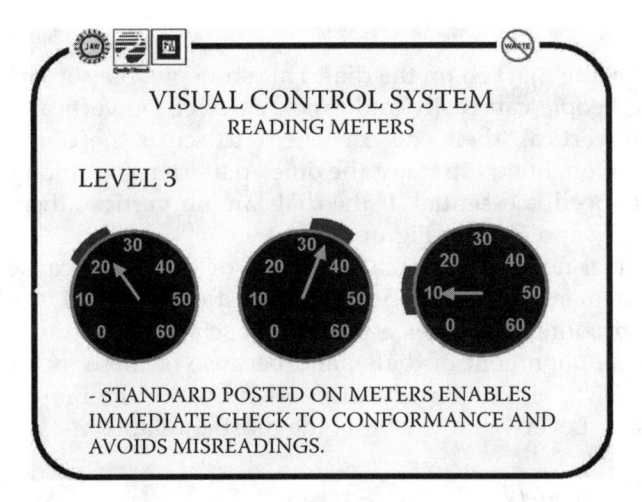

Figure 12.7 (b) This scheme has less common cause variability than Level 1, but still leaves room for potential errors due to judgments about pointer position.

walk to the meter and then back to the reference book. This is also a non-value-added activity.

The Level 3 strategy is less error-prone and exhibits less inherent variability. Here, the acceptable region is marked on the meter dial. The operator simply notes if the meter readings fall within the marked region. Judgments arise if the meter dial falls on a boundary. See Figure 12.7(b).

The Level 4 strategy reduces the cognitive load and hence inherent variability of the task even further. The meter dials are rotated so that for the acceptable readings, all the dials are oriented vertically, the acceptable

VISUAL CONTROL SYSTEM
READING METERS

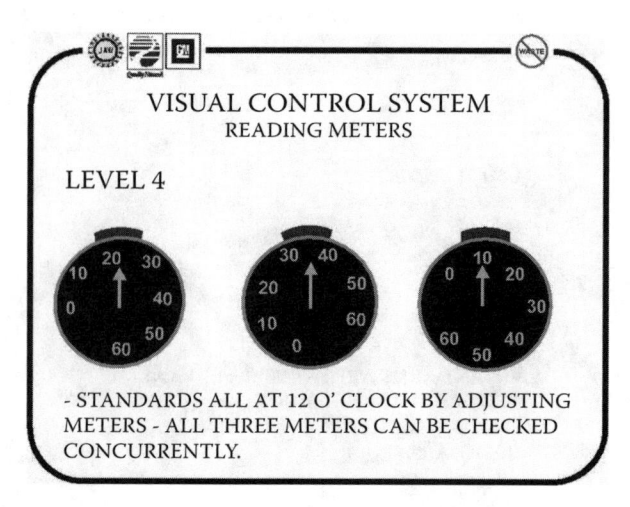

LEVEL 4

- STANDARDS ALL AT 12 O' CLOCK BY ADJUSTING METERS - ALL THREE METERS CAN BE CHECKED CONCURRENTLY.

Figure 12.7 (c) The vertical orientation of the correct readings increase reliability and reduces common cause variability if speed is important (e.g., dashboard dials in a racing car).

regions still being marked on the dial. This strategy relies on human pattern recognition. People can easily and quickly detect the vertical alignment. If a dial is off vertical, then one can check to see if the dial is within an acceptable region. In this strategy, the quick pattern recognition scheme adds reliability if speed is essential. If the dials are off vertical, then the marked region strategy applies. See Figure 12.7(c).

A universal design approach would favor Level 4 because it is cognitively least demanding, is ergonomically least demanding, is efficient, provides error-proofing, and uses a perceptual scheme that humans are good at (the vertical alignment of dials); and because of these factors, it has the least inherent variability. Finally, all workers would be more efficient and accurate using Level 4 and thus it is the most equitable.

Example 2: Airport

Airports use a variety of techniques to reduce common cause variability for passengers and customers using their facilities. Figure 12.8 shows an airport terminal lobby. The lobby is clean and organized following the 5S strategy. Visual controls, the signage, help to control and manage the flow of passengers through the terminal. Figure 12.9 shows a representation of the terminal's environmental organization. Ticketing, check-in, and security functions follow highly structured, standardized operating procedures. When these standardized procedures are clearly communicated to both employees and passengers, the process works relatively smoothly.

Note the high contrast between the signage letters in Figure 12.8 and the use of a simple, large font style for lettering. These features ensure

Chapter twelve: Stable and predictable 173

Figure 12.8 The interior signage and layout of a modern airport terminal lobby. This image illustrates a variety of visual control, workplace organization, and standardized operating procedures.

adequate perception of the signage. Not shown in the figures is a color-coding scheme used to designate different floor levels. Also not shown are the error-proofing strategies built into the e-ticketing system. Under program control, a device requests and accepts ticketing and passenger data; it verifies the information via a Web-based procedure and prints a boarding pass. The elements of the e-ticketing system exhibit device and process reliability, and quality of service associated with sound engineering design practices.

Figure 12.9 is used as a visual control; it builds knowledge into the terminal environment, to inform passengers of their location and the location of other facilities and services. It supports the workplace organization and helps create more of a self-managing system with respect to passenger navigation.

Airport terminals must utilize a universal design approach to its facilities and passenger services. The facilities and services must be accessible as mandated by the ADA and Access Board guidelines. The facilities and services must handle people of all sizes, all nationalities, with widely ranging reading and language skills. By and large, most major airport terminals functional very well.

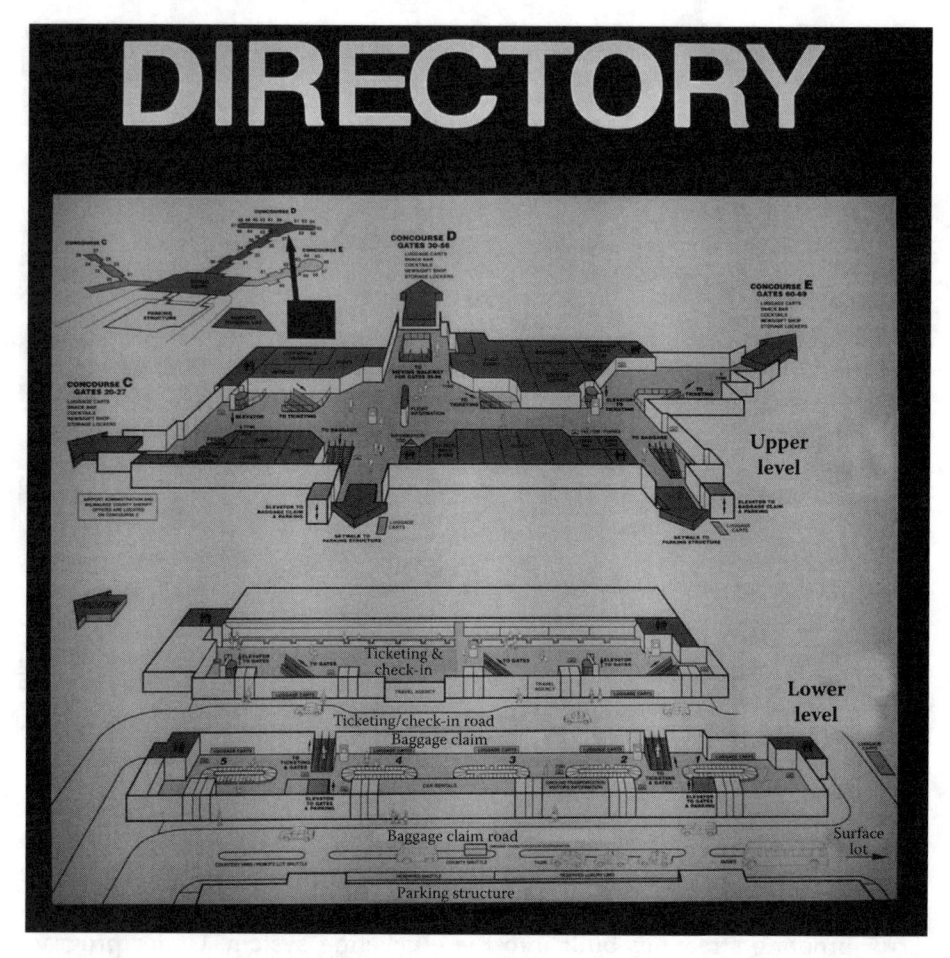

Figure 12.9 This image is used as a visual control to build knowledge into the terminal environment. It is a representation of the terminal layout.

Conclusions

The systematic application of all the previously discussed universal design principles yields powerful synergistic effects, the most pronounced being the reduction in common cause variability. The reduction in common cause variability renders designed entities more accessible and usable by a larger number of people.

Designers cannot rely on a user possessing high skill levels or expertise to use the designed entity. Recall the broom balancing example from earlier in this chapter; or the absolute judgment experiments discussed in Chapter 7, which demonstrated that most people can make correct judgments for about six (the magic number seven plus or minus two) different perceptual inputs [50]. However, it was also noted that a person with perfect pitch can often distinguish greater than 50 pitches correctly [50]. Hence, while people can adapt and use their skills and expertise to overcome poor design strategies, such designs limit the accessibility and usability of designed products, processes, jobs, and services.

> If the system is stable—with only predictable and acceptable variation—it can be continually improved to generate increased productivity. If the system isn't stable, there's no point in setting a goal. Only frustration and lower morale will result [29].

References

1. W. E. Deming, *Out of Crisis*, 15th ed. Cambridge, MA: Massachusetts Institute of Technology, 1982.
2. R. Reich, *The Future of Success*. New York: Alfred A. Knopf, 2000.
3. NIST, NIST at 100: Foundations for Progress, National Institute of Standards and Technology, 2002.
4. ANSI, American National Standards Institute, http://www.100.nist.gov/ANSI, 2004.
5. ISO, International Organization for Standardization, http://www.iso.org/iso/en/ISOOnline.frontpage, 2006.
6. Institute of Electrical and Electronics Engineers, IEEE 1220-2005 IEEE Standard for Application and Management of the Systems Engineering Process, http://www.techstreet.com/cgi-bin/detail?product_id=1260785, 2005.
7. INCOSE, The International Council on Systems Engineering-Home web page, http://www.incose.org/, 2006.
8. M. Pecht and A. Ramakrishnan, Development and Activities of the IEEE Reliability Standards Group, *J. Reliability Eng. Assoc. Jpn.*, 22, 699–706, 2000.
9. M. Levin, A. and T. Kalal, T., *Improving Product Reliability: Strategies and Implementation*. New York: John Wiley & Sons, 2003.
10. Fairchild Semiconductor, Fairchild Semiconductor Product Reliability Report, http://www.fairchildsemi.com/company/quality/Reliability_report.xls, 2006.
11. T. Ohno, *Toyota Production System: Beyond Large Scale Production*. Portland, OR: Productivity Press, 1988.
12. T. Pyzdek, Motorola's Six Sigma Program, *Quality Digest*: http://www.qualitydigest.com/dec97/html/motsix.html, 1997.
13. FDA, FDA Announces Actions to Strengthen its Postmarket Program for Medical Devices Effort Will Improve Management of Adverse Events, Enhancing Patient Safety, http://www.fda.gov/bbs/topics/NEWS/2006/NEW01506.html, 2006.
14. Underwriters Laboratories, Underwriters Laboratories Homepage, http://www.ul.com/, 2006.
15. IEEE, The IEEE Reliability Society, http://www.ieee.org/organizations/rab/join/RL007.html, 2006.
16. J. Won, D. Cochran, H. T. Johnson, S. Bouzehouk, and B. Masha, Rationalizing the Design of the Toyota Production System: A Comparison of Two Approaches, presented at *International CIRP Design Seminar Proceedings*, Stockholm, 2001.

17. The Institute of Industrial Engineers, IIEnet.org: Your new home page, http://www.iienet2.org/Default.aspx, 2006.
18. SME, Society of Manufacturing Engineers Home Web Page, http://www.sme.org/cgi-bin/communities.pl?/communities/techgroups/nano/nano_mems_hp.htm&&&SME&, 2006.
19. ASQ, American Society for Quality, http://www.asq.org/, 2006.
20. P. H. Barringer, Process Reliability and Six-Sigma, presented at *National Manufacturing Week Conference*, Chicago, IL, 2000.
21. Wikipedia®, Six Sigma, http://en.wikipedia.org/wiki/Six_Sigma, 2006.
22. A. Kiran and C. Kaplan, Six Sigma — It's not just for manufacturers anymore, http://ad.doubleclick.net/click%3Bh=v8/3474/3/0/%2a/v%3B47629065%3B1-0%3B0%3B14168657%3B3454-728/90%3B18630048/18647943/1%3B%3B%7Efdr%3D51293602%3B0-0%3B0%3B10365707%3B3454-728/90%3B18510317/18528212/1%3B%3B%7Esscs%3D%3fhttp://www.sap.com/mk/get/ussp_bantech, 2004.
23. D. Jones, Taking the Six Sigma approach, http://www.usatoday.com/money/companies/management/2002-10-30-sigside_x.htm, 2002.
24. Q. Tools, Quality Tools is a clearinghouse for practical, ready-to-use tools for measuring and improving the quality of health care. QualityTools is sponsored by the Agency for Healthcare Research and Quality (AHRQ), http://www.qualitytools.ahrq.gov/, 2006.
25. M. H. Jones, Six-Sigma — at a Bank?, *Six Sigma Forum Magazine*, 2004, pp. 13–17.
26. Banker's Academy, Managing Banking Operations for Productivity & Quality, http://www.bankersacademy.com/managingops.php, 2006.
27. L. S. Pheng and M. S. Hui, Implementing and Applying Six Sigma in Construction, *Journal of Construction Engineering and Management*, 130, 482–489, 2004.
28. iSixSigma LLC, Six Sigma in the Insurance Industry: On line threaded discussion of the iSixSigma LLC, http://main.isixsigma.com/forum/showmessage.asp?messageID=2007, 2006.
29. J. Pryor, Deming's Point #11 as Applied to the Insurance Industry, http://www.irmi.com/Expert/Articles/2006/Pryor09.aspx, 2006.
30. J. Reese, Army Adopting Lean Six Sigma, *Army News Service*, http://www.military.com/features/0,15240,87414,00.html, 2006.
31. E. McClanahan and C. Wicks, *Future Force: A Teacher's Handbook for Using TQM in the Classroom*. Chino Hills, CA: PACT Publishing, 1993.
32. MAEF, *Classroom Quality Program*. St. Paul, MN: Minnesota Academic Excellence Foundation, 1996.
33. C. Wicks, J. Peregoy, and J. Wheeler, *Plugged In! A Teacher's Handbook for Using Total Quality Tools to Help Kids Conquer the Curriculum*. New Bern, NC: Class Action, Coast to Coast Connection, 2001.
34. D. Mehrotra, Six Sigma in Education, http://www.isixsigma.com/library/content/c011029a.asp, 2006.
35. W. R. Miller, The Definition of Design, http://www.tcdc.com/dphils/dphil1.htm, 1996.
36. M. Imai, *Kaizen*, 1st ed. New York: McGraw-Hill, 1986.
37. M. Imai, *Gemba Kaizen*. New York: McGraw-Hill, 1997.
38. J. Ishiwata, *I.E. for the Shop Floor 1: Productivity through Process Analysis*. Cambridge, MA: Productivity Press, 1991.

39. Japan Human Relations Association, Kaizen Teian 2, J. H. R. Association, Ed., First English ed. Portland, OR: Productivity Press, 1992.
40. W. Wanyama, A. Ertas, H.-C. Zhang, and S. Ekwaro-Osire, Life-Cycle Engineering: Issues, Tools AND Research, *Int. J. Computer Integrated Manufacturing*, 16, 307–316, 2003.
41. Dassault Systemes, Dassault Systemes Home Web Page, http://www.3ds.com/index.php, 2006.
42. COE, COE is an independent professional organization run by users, for users, with the support of Dassault Systemes & IBM Product Lifecycle Management, http://www.coe.org/, 2006.
43. S. Shingo, *A Study of the Toyota Production System*. Portland, OR: Productivity Press, 1989.
44. J. P. Womack, D. T. Jones, and D. Roos, *The Machine That Changed the World*. New York: Harper-Collins Publishers, 1990.
45. J. P. Womack and D. T. Jones, *Lean Thinking*. New York: Simon & Schuster, 1996.
46. R. Erlandson, Business and Legal Conditions Supporting the Employment of Individuals with Disabilities, in *The Sourcebook of Rehabilitation and Mental Health Practice*, D. Moxley and J. Finch, Eds. New York: Plenum Press, 2003, pp. 51–60.
47. R. F. Erlandson, Accessible Design and Employment of People with Disabilities, in *The Sourcebook of Rehabilitation and Mental Health Practice*, D. Moxley and J. Finch, Eds. New York: Plenum Press, 2003, pp. 235–252.
48. Perry. D., History of Karate Kata, PageWise, Inc: http://vtvt.essortment.com/karatekata_rfbg.htm, 2002.
49. S. Spear and H. Bowen, Decoding the DNA of the Toyota Production System, *Harvard Bus. Rev.*, September-October, 96–106, 1999.
50. G. Miller, The Magical Number Seven, Plus or Minus Two: Some Limits on Our Capacity for Processing Information, *Psycholog. Rev.*, 63, 81–97, 1956.

chapter thirteen

Equitable: Transcending, integrating

Chapter goals:

- State the equitable universal design principle.
- Provide a rationale for its relevance.
- Provide design strategies that work to address the principle.
- Provide example applications of the strategies.
- State the synergistic, transcending and integrating features of the principle.

Equitable: Transcending, integrating

Principle

Universally designed entities should be equitable; that is, the entities should provide the same means of use for all users: identical whenever possible and equivalent when not possible. The products and processes should avoid segregating or stigmatizing any users, making the design appealing to all users [1].

Discussion

Equitability transcends or goes beyond the other universal design principles in that if it is satisfied, it integrates elements from all the other principles and exhibits features and characteristics that are unique to being equitable. Equitability imposes constraints on the other design principles in that they must be applied so that the designed entities are accepted by a broad spectrum of users. The designs must be age and context appropriate as well as aesthetically pleasing. These design features and characteristics are true for any product or service, but are especially true for designers seeking to follow universal design principles.

Accessible design does not concern itself with being equitable. Rather, accessible design is concerned with removing or preventing environmental barriers as prescribed by law, guidelines, and standards. This does not mean that an accessible design requirement cannot be equitable.

Design strategies

The following design strategies are presented without a lot of discussion. Each of these strategies, while simply stated, could fill a book in their own right. The important thing is that designers are aware of and sensitive to these issues.

Strategy 21: Design entities that are age and context appropriate

The designed entities must be age and context appropriate, or their use will draw undue attention and people simply will not use those entities.

Curb cuts are probably the classic example of equitable universal design. People riding bicycles, people pushing baby strollers, elderly people who have trouble ascending/descending stairs, and people in wheelchairs all use the curb cuts. There is no discrimination. Everyone is better served. Automatic door openers on buildings and public facilities also exemplify an equitable design. Based on sensor activation, a door opens for everyone: people whose arms are full, people who are engaged in conversation, people who are in wheelchairs, and the elderly with walkers.

The adjustability features on cars (adjustable seat, steering wheel, floor pedals, and mirrors) allow a variety of users to customize the driver's seat and control access to their individual needs.

Finally, work tools and processes should provide the same means of use for all users: identical whenever possible, equivalent when not possible. For example, counter-balances and positioning brackets are used to counter the weight of a drill and position its angle of attack such that the worker can focus attention on the task as opposed to holding the drill while having to carefully aim and position it to do the job. Such designs are ergonomically sound, reduce the common cause variability of the task, and are beneficial to all workers.

Strategy 22: Design entities that are aesthetically pleasing

If the entity is not aesthetically pleasing it will not be commonly accepted and utilized.

Figures 13.1(a) through (d) show a collection of devices manufactured by OXO International, Ltd. [2]. OXO has embraced universal design as its corporate design philosophy and has been extremely successful with the sale of its products along with industry and artistic recognition of their products' functionality, aesthetically pleasing designs, and competitive pricing [3]. As

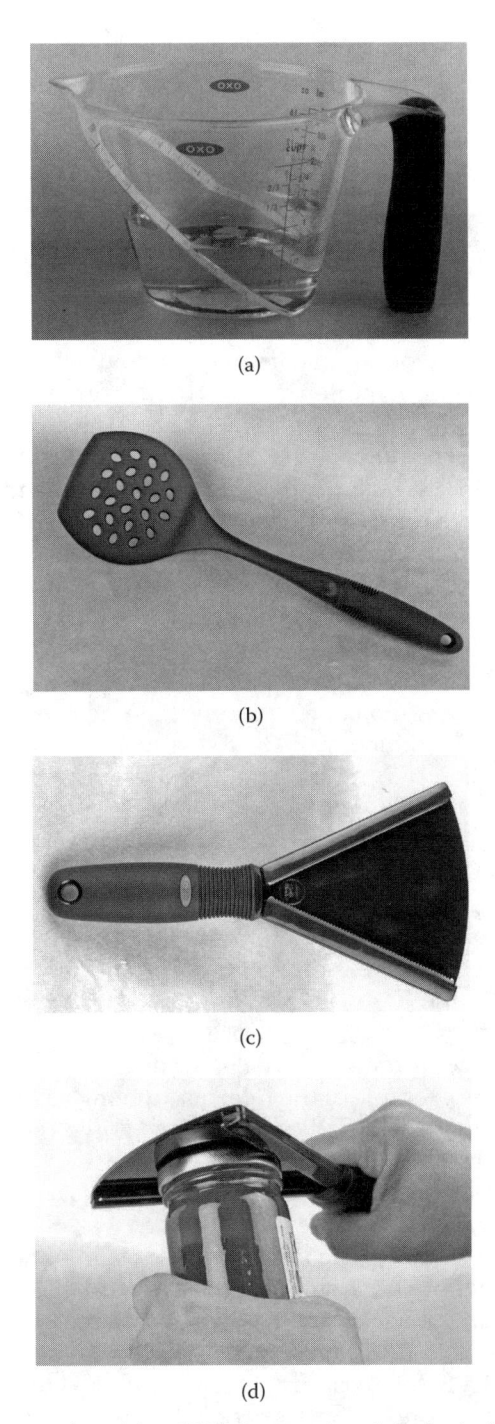

Chapter thirteen: Equitable: Transcending, integrating 181

Figure 13.1 Devices produced by OXO International, Ltd., include: (a) measuring cup with measuring scale viewable by looking in from the top, (b) a spatula, (c) a jar opener, and (d) the jar opener being used.

noted in Chapter 1 a collection of OXO products are part of a permanent Design Collection of The Museum of Modern Art, New York [3].

OXO International, Ltd., and its product line clearly demonstrate that products designed using universal design principles can be not only functional, but also aesthetically pleasing and competitively priced.

Strategy 23: Design entities that are competitively priced

If the entity is too expensive, it will not be widely accepted or used.

It is a common trend for new technology to start out being very expensive and available only on luxury items; and then as companies gain engineering and fabrication experience, the cost of providing the technology drops and finds its way into more and more products, thereby reducing costs even further. GPS technology is going through this cost-reduction process and thereby rendering global positioning, navigation, and object or person detection technologies much more affordable and accessible to all people. A GPS device can help us find our way in a strange city, and it can also be used to locate a person with Alzheimer's who has wandered away from home and has a GPS-equipped mobile phone or customized sensing device. GPS devices can help cognitively impaired people navigate and become more independent.

Another trendy, age and context appropriate, aesthetically pleasing, and competitively priced product available to everyone, including people recovering from traumatic brain injury, stroke, or other memory-compromising conditions is "GPS Toes." Quoting from their advertising material; "Using similar low-power, nano-derived technology, GPS Toes are toe rings that communicate to a GPS receiver kept in a purse or worn on a belt. Wearing one on each foot, the GPS Toes device will guide the wearer to a preset destination by vibrating and lighting up to signal upcoming direction changes. The left toe ring will indicate left turns and the right toe right turns, whether driving on the highway, walking on city streets, or hiking on the mountain trail" [4].

The universal accessibility of eyeglasses and corrective lenses exemplifies a more mundane example of the interplay among technology, cost reduction, and social values. It took the convergence of several threads of technology and changing societal values before eyeglasses and corrective lenses could be considered truly affordable and accessible for everyone. The ability to manufacture high-quality lenses from plastic was one of the major technological breakthroughs [5]. The advent of eye examination equipment that allowed a determination of the shape of the eye and hence a way to prescribe the design and shape of the corrective lenses was the other major technical advance [5]. Being able to design, by prescription, a corrective lens for an individual at relatively low cost dramatically reduced the cost of providing eyeglasses, allowing more and more people to buy eyeglasses and corrective lenses. The low cost of providing corrective glasses and lenses, along with the functional improvement of people using the devices, radically changed

Chapter thirteen: Equitable: Transcending, integrating   183

Figure 13.2 Electronic devices such as cameras and binoculars now come equipped with visual stabilization circuitry using embedded microprocessors.

society's concept of disability due to visual impairments. The final step in ensuring broad accessibility was a change in social values that led to eyeglasses and corrective lenses being covered by medical insurance plans [5]. The universal availability of $5 to $25 "reading" glasses in supermarkets, bookstores, pharmacies, and the like speaks to the power of the baby boomers to influence markets.

Technology transfer is another way that new technologies find their way into the broader consumer market. Helmet-mounted, heads-up target stabilization systems for attack helicopter and fighter pilots have evolved into visual stabilization systems for cameras and binoculars. Cameras and binoculars now routinely contain microprocessor technology that stabilizes the image for all users. This allows a tired person, or a person with tremors or other motor problems, to use the same device. Introducing such image stabilizing technology into a broad range of consumer products, such as the digital camera shown in Figure 13.2, has enabled the cost of providing such technology to decrease dramatically and thereby render the feature more affordable to everyone [6, 7].

Government agencies such as the Department of Defense (DOD) and NASA have strong technology transfer programs. The DOD program is called DOD TechMatch [8]. NASA has a Technology Opportunities website [9]. The following quote from the NASA website is typical of the government technology transfer program.

> U.S. businesses can take advantage of a vast pool of NASA-developed technology resources to find solutions to technical problems. Through a program known as Technology Transfer, NASA technology can be adapted, at little or no cost, to meet a particular

need—saving your company valuable time and money. Applications for NASA technology are widespread, from building prototypes of new production items to nondestructive testing of an existing component [9].

An example of a NASA technology is the "Headset Assembly (for Behind-the-Ear Hearing Aids)" developed at the Marshall Space Flight Center (MSFC). This patented headset helps eliminate the high-pitched feedback noise associated with behind-the-ear hearing aids [9].

Strategy 24: Market the entity for as broad a demographic and socioeconomic base as reasonable and possible

If the consuming public perceives the entity as being for the disabled community, the entity will not find widespread acceptance.

Closed captioning, while starting out as an accommodation for the deaf and hard-of-hearing, has become quite common, emphasizing the fact that everyone experiences hearing difficulties at some time or in some environments. Restaurants, bars, health clubs, and information monitors in public facilities typically use closed captioning on their public televisions or information monitors.

The AlphaSmart© is a small, portable, inexpensive computer dedicated to note taking, i.e., an electronic appliance. When the AlphaSmart© was first developed and hit the market it was seen by many people to be targeted for individuals with cognitive impairments; those who needed organizational help, who found computers too complicated, i.e., people with traumatic brain injury, or various learning disabilities. This perception limited early sales.

Figure 13.3 AlphaSmart©: while still extensively used in special education settings, it has become a trendy note-taking device for all kinds of students, from elementary school through college.

A transition occurred in the market place as people began to see its broader market potential. It is currently being marketed as a study aid and has found wide application in schools. Elementary schools use the device for developing basic writing skills and keyboard skills. High school and college students use the device for note taking and organizational aids. Today AlphaSmarts© are "cool," students like them and are eager to use them.

Transcending and integrating

Equitability is discussed last because it is a value judgment that universal design should be equitable. As such, the requirement for equitability places constraints on and forces the integration of the other universal design principles and strategies. For this reason, equitability sits at the top of the universal design hierarchy.

The designed entities should be equitable in that the entities should provide the same means of use for all users: identical whenever possible and equivalent when not possible [10]. The products and processes should avoid segregating or stigmatizing any users, thus making the design appealing to all users. The products used as examples throughout this book, such as building entrance ramps, automatic door openers, the Windows and Apple operating systems that include accessibility options as a standard part of the system, public kiosks that provide wheelchair access, and both auditory and visual user I/O are all examples of equitable designs that avoid stigmatizing users. These examples provide evidence that the equitable design of products and processes is not only possible, but also not necessarily costly in terms of funding, time, and resources.

These examples also illustrate that building equity into the design process is not an afterthought, and that equity must be an essential part of the problem statement and definition. It must be an explicit part of the design requirements. If this is the case, it requires that universal design must be an explicit design strategy and the universal design principles must be explicitly addressed from the beginning of the design process through to the end.

Therefore, while equity is the final universal design principle considered in this discussion, its representation as the integrating force of the other universal design principles argues for its prominence in design requirements. This is why it is often the first-cited universal design principle in other discussions.

References

1. B. R. Connell, M. Jones, R. Mace, J. Mueller, A. Mullick, E. Ostroff, J. Sanford, E. Steinfeld, M. Story, and G. Vanderheiden, Principles of Universal Design, http://www.design.ncsu.edu:8120/cud/univ_design/princ_overview.htm, 1997.
2. OXO, What We're About, http://www.oxo.com/about_what.php, 2004.

3. OXO, List of Awards for OXO Products, http://www.oxo.com/about_awards.php, 2004.
4. textually.org, Technojewelry — Penta Phone, Ring Phone and GPS Toes, a website featuring exotic and novel GPS devices: http://www.textually.org/textually/archives/2005/05/008354.htm, 2005.
5. M. L. Rubin and G. M. Hope, Optics and Refraction, *Ophthalmology*, 103, S102–S108, 1996.
6. P. Askey, Minolta DiMage A1 Review, http://www.dpreview.com/reviews/minoltadimagea1/, 2003.
7. VMI Inc, Promotioinal Ad for a Sony FCBEX980S Camera with Image Stabilization Technology, http://catalogs.infocommiq.com/avcat/CTL1417/index.cfm/rcs_id/1417/NOTRACK/1/ProdID/331430/T3/1109.htm, 2006.
8. DOD, DOD TechMatch, a Web-based portal designed to provide industry and academia a Department of Defense-sponsored solution to find research & development opportunities, licensable patents, and information on approximately 120 DoD labs located across the United States, http://www.dodtechmatch.com/DOD/index.aspx, 2006.
9. NASA, Technology Opportunities, http://techtran.msfc.nasa.gov/techopps.html, 2006.
10. M. F. Story, Maximizing Usability: The Principles of Universal Design, *Assistive Technol.*, 10, 4–12, 1998.

chapter fourteen

Universal and accessible design in the workplace

Chapter goals:

- Describe properties of the workplace that influence the application of universal design.
- Present examples of the application of universal design for large multinational corporations, a small business, and a not-for-profit.
- Discuss how different management strategies and decisions influence the outcomes associated with the application of universal design in different business workplaces.

Consumer applications versus workplace applications

The application of universal and accessible design to the workplace starts with the design of consumer products and services. From a customer product and service perspective, the design objectives include functional requirements, selling price, and potentially conflicting demands—for example, between the amount and kinds of flexibility afforded versus operational complexity. Depending on the product or service, there may be legal mandates for accessibility for buildings, telecommunication products and services, and electronic and information technology. Furthermore, the universal and accessible design requirements must be balanced against a host of other design requirements, such as any standards (ISO, ANSI, etc.), designing for cost-effective manufacturing and assembly, as well as service and maintenance once in the field, and then end-of-life recycling or disposal.

It is the process of designing a product for manufacturing and assembly that the design of the workplace environment begins. So interdependent is the design of the product and its manufacture that companies such as Dassault Systemes have developed CAD software, such as CATIA and DELMIA, that contains tools that automatically design the manufacturing or assembly fixtures as the product itself is being designed (fixtures are the devices that hold the product being manufactured or assembled as it moves from workstation to workstation) [1]. Fixtures, the assembly workstations, and the assembly line itself are examples of what needs to be designed within the workplace.

In Chapter 2 universal design was defined as the design of entities that can be used and experienced by people of all abilities, to the greatest extent possible, without adaptations. When considering the workplace, "entities" can mean the physical plant (for example, parking, entrances, aisle ways, toilets, workcells, assembly lines, stairs, ramps, etc.); the machines and tools that workers use (for example, conveyors, drill presses, hand-tools, computers, check-out systems, bar-code scanners, etc.); the work processes (for example, assembly, manufacturing, forms, re-cycling, servicing, etc.); and the organization of the work (for example, small teams, lean production, total quality management, just-in-time, break schedules, assembly rates and cycle times, downsizing, reengineering, etc.).

The organizational context—such factors as management structures, supervisory practices, and human resource polices—determine what and how things happen in the workplace, that is, on the shop floor, in the assembly area, the retail floor, the restaurant, or the office. In turn, external pressures and forces—such as economic developments, trade policies, national and international standards, technological innovations, changing worker demographics, and the supply of labor—all influence the organizational elements and decision making. All in all, the workplace is a very complex environment for the application of universal and accessible design principles.

Implementation of universal design in the workplace: Examples

Five examples are presented illustrating the application of universal design in the workplace. These examples range from large multinational corporations in the automotive (Subaru), retail (Wal-Mart), and fast food (McDonald's) industries to a relatively new small business, typical of today's global economy, producing hand-painted home and garden tools (Art for a Cause), and finally a not-for-profit service organization (Goodwill Industries). These examples exemplify a broad range of organizational contexts, size and scale of operations, and external pressures.

Fuji Heavy Industries Ltd.

The first example considers the application of universal design elements to Fuji Heavy Industries, Gunma Plant, manufacturing site for Subaru vehicles [2]. The Gunma Plant utilizes lean production, TQM, and *kaizen* methods [2]. It is within this context that the "Universal Plant Project" was initiated. The following quote lays out the rationale for the project and specifies its goal.

> Creativity is required in a manufacturing industry and Subaru is working hard to create a workplace where the attributes of all employees can be harnessed to enhance our company's creativity. Subaru wants a diverse work force. To accomplish this goal, we are making efforts to build a barrier free workplace. Barrier free does not only refer to the physical structures in the workplace, but also the mindset of all workers. Barriers don't just prevent the disabled from contributing fully to a company, they also prevent women and the elderly from contributing fully. At Subaru, we are continually making efforts to eliminate all barriers by applying universal design elements, not only to our products, but to our workplace as well [2].

From the project's description it is clear that establishing a diverse workforce and addressing the needs of people with disabilities were the major driving forces. The project had four initiatives: recruit disabled staff, remove concrete barriers, remove mindset barriers, and inspiration [2]. People with disabilities but with the requisite education and skills were recruited and became part of the planning team. The expression "Remove Concrete Barriers" was a descriptive term for removing physical barriers within the plant. Examples of removing physical barriers included; making the plant wheelchair accessible, providing accessible parking places, installing different height lockers and more adjustable furniture. Removing physical barriers also meant creating a more universally designed workplace, e.g., adjustable worktables, mobile and adjustable inventory storage for assembly areas, conveyors to eliminate carrying and lifting. Consistent with the application of universal design principles the removal of these "Concrete Barriers" yielded benefits for all workers, i.e., a safer workplace and higher quality products [2].

The "remove mindset barriers" initiative was designed to provide support to workers with disabilities while demonstrating that such support enabled those workers to perform their jobs competently. The initiative included seminars, for all employees from management on down, on universal and accessible design, testimony and data from workers demonstrating the enabling effects of the technology and redesign on job creation and enhanced performance, and sign language classes for supervisors [2]. The final initiative was called "inspiration" and involved such things as optional sign language classes for employees, and idea contests where all workers were invited to present ideas for accessible and universal design innovations and changes. Of course, a *kaizen* process is used to ensure continuous improvement with respect to the application of universal design in the plant.

How is this project unfolding?

> Through these activities we were able to double our disabled work force, from 63 people five years ago to 112 people now. Barrier free not only refers to the infrastructure, but also to the

mindset of all staff. In the beginning we were mostly concerned about the concrete structures, but now our focus is changing to the mindset of all staff. In the beginning we thought that hiring disabled staff mostly required money, but now we realize that hiring disabled staff leads to enlightenment. To improve the mindset companywide, we are making efforts to not put all disabled people in one work area, but to add them to sections throughout the company.

Applying universal design to the workplace has enhanced our company's creativity, made our workplace more diverse and has made the working atmosphere much more lively. [2]

This example presents a conscious management decision to effect social as well as business change. The next two examples illustrate different management decisions.

Retail and fast food industries

This section considers the retail industry, Wal-Mart, and the fast food industry, McDonald's. In both of these examples, companies innovated, explored, made use of, and facilitated the widespread use of what we are calling universal design principles, methods and strategies without explicitly recognizing them as such. In both cases, the companies embraced lean production, TQM, and Six Sigma methods to reduce costs, reduce non-value-added activities (NVAAs), and thereby reduce process variability and maintain quality. In both cases, the companies focused on providing customers products and services at low prices in a pleasant, friendly environment. In both cases, the companies came to dominate their respective industries and promote or force change onto their respective industries and supplier networks.

Retail: Wal-Mart

As reported by Dicker, in his book entitled *The United States of Wal-Mart*, Wal-Mart is the largest corporation in the world, the largest private employer in the United States, and owner of the nation's largest private trucking fleet [3]. Wal-Mart focuses on keeping its costs low and generating high-volume sales. Wal-Mart is basically a retail operation but with a growing mix of associated business interests such as real estate [3]. The various Wal-Mart stores are designed to attract customers and sell products and services.

When Sam Walton, Wal-Mart's founder, started his first retail operations, his main concern was keeping prices low. His early stores were reportedly a mess, with clothes strewn around on tables, open boxes, and unpredictable inventory [3]. When Walton started his retail businesses, customers were serviced by clerks who helped customers find merchandise or actually retrieved it from the storeroom for the customer. Walton heard of a small retailer who allowed customers to move freely among stock organized into displays around the store and actually pick up their own merchandise and then move to a checkout counter to pay. Walton thought this was ingenious; it cut labor costs dramatically and opened up more floor space for merchandise display and sale [3].

In terms of universal design, what Walton was creating was a self-managing retail environment. That concept has grown into a very sophisticated design and operational strategy for all Wal-Mart retail stores. Technology has allowed Wal-Mart to create one of, if not the best inventory management, tracking, and distribution systems in the world [3]. This is the ultimate self-managed environment utilizing knowledge built into the environment, strict standardized operating procedures, and workplace organization to the max. Massive computer databases, communications technology, bar codes, and now radio frequency identification (RFID) tags enable Wal-Mart to track inventory and keep merchandise on the shelves specific to local customer demands like no other retail operations has ever done [3].

In terms of its own operations, Wal-Mart has embraced Six Sigma (in a relentless process of removing non-value-added activity), *kaizen*, and lean production strategies such as 5S, error-proofing, standardized operating procedures, workplace organization, visual controls, and just-in-time inventory practices. The retail sales floor is the key stage. It must exhibit all the best practices of a self-managing process, but with a human touch, happy greeters and associates, who make sure the customer has a good shopping experience.

The design of the products that Wal-Mart sells is basically the responsibility of Wal-Mart's suppliers, with one major exception: cost. Wal-Mart's size and purchasing power have enabled it to dictate what it will pay, what kind of cost reductions it expects, and what level of quality it expects [3]. Wal-Mart's relentless cost-cutting pressures have also forced its suppliers to adopt the same Six Sigma, *kaizen*, lean production, and just-in-time inventory practices.

Fast Food: McDonald's

In *Fast Food Nation*, Schlosser looks at the fast food industry and the impact it has had on the meatpacking industry and agriculture [4]. In 1948, the McDonald brothers originated the "Speedee Service System" for fast food preparation and service. In their ad seeking franchisees, the brothers describe the benefits of their new system:

> Imagine – No Carhops – No Waitresses – No Dishwashers – No Bus Boys – The McDonald's System is Self Service! [4].

In the mid-1950s, Ray Kroc was selling milk-shake mixers and he wanted to find out why the new McDonald's Self-Service Restaurant in San Bernardino, California, needed eight of his five-at-a-time mixers. Kroc was amazed with the self-service operation and was able to "purchase the right to franchise McDonald's nationwide" [4].

Kroc experimented with and perfected a collection of strategies to further improve the efficiency of the original self-service. McDonald's and other fast food companies must attract and serve customers. McDonald's utilized many *kaizen* and universal design strategies—before they were called that—because they made sense. In universal design terms, McDonald's used standardized operations and workplace organization, 5S, visual controls, and quality controls to ensure a consistent level (low variability) of food and service quality, whatever McDonald's outlet you went into wherever you were [4]. Other fast food operations were quick to follow suit.

As McDonald's and other fast food companies grew, they made demands on their suppliers for cost reductions. In response to these demands and other global pressures, McDonald's suppliers (for example, meatpacking companies and potato suppliers) had to reduce costs. These suppliers were forced to invest in new technologies and also adopted many of the Six Sigma, lean production strategies [4].

Comparisons: Wal-Mart and McDonald's

The patterns for both Wal-Mart and McDonald's are similar. Both pioneered, explored, experimented with, and successfully adopted methods and strategies that we are collecting under the umbrella term "universal design." Both grew to dominate their respective industries and both have forced their competitors and supply-chain businesses to adopt similar methods. Both companies share another common feature, that of labor problems [3–5].

The fast food industry and Wal-Mart experience very high hourly paid employee turnover rates [3, 4]. The reasons for this are similar: low pay, no or very expensive health care benefits, questionable use of immigrant laborers, and questionable labor practices [3–5]. The point of this discussion is not to explore the social, economic, or political implications of these practices, but rather to note that universal and accessible design principles and methods can support such employment practices.

A small company: Art for a Cause

Art for a Cause, LLC, started in 1998 by Lisa Knoppe-Reed, is a small business in Birmingham, Michigan [6, 7]. CuteTools™ are the products of Art for a Cause, LLC. What cause? The employment of people with disabilities, providing them the dignity of work and a sense of self-worth.

CuteTools™ is a collection of more than 30 kitchen, garden, and home tools. They are cute because of the decorative patterns hand-painted on their wooden handles and because of the signature ribbon tied in a bow around the handle. Folks with disabilities perform the sanding and paint priming jobs. In addition to the people who come to work at the Art for a Cause facility, sanding and priming kits are sent to special education schools around the country where students sand, prime, and return the prepared tools for final painting. Trained artists at Art for a Cause paint a decorative pattern on the prepared handles. Workers with disabilities then help to prepare and affix the ribbons, prepare the shipping boxes, package the finished products, and ship them.

Product sales are steadily growing. In November 2005, CuteTools™ were shipped to 350 stores, mostly in the United States. As of September 2006, CuteTools™ were being shipped to more than 2500 stores worldwide [7]. As astounding as this growth are the number of people who want to replicate this model in their own states and countries. Talks are under way with potential partners and collaborators in five states, as well as several countries, Canada, Mexico, Ireland, Scotland, and Great Britain [7]. While participating in a trade show in Tokyo, Japan, in July 2006, Knoppe-Reed was approached about starting a business operation in Japan. A Tokyo facility was opened in October 2006. Easter Seals International is working with Art for a Cause to provide workers in Japan, along with students from regional vocational and rehabilitation training organizations.

Art for a Cause exemplifies the new e-commerce global economy. The Wayne State University Enabling Technologies Laboratory (ETL) has been working closely with Art for a Cause for more than 2 years [6]. Student design projects are targeting the design and implementation of accessible and enabling technology based on accessible and universal design principles so as to allow Art for a Cause to continue to employ people with disabilities and meet the growing production demands while maintaining quality products [8].

Small retail operations: Goodwill Industries

Goodwill Industries of Southeastern Wisconsin, Inc., has embraced *kaizen* and quality control methods, and each of its operations is a model of *kaizen* approaches and successes. "Goodwill's mission is to provide work opportunities and skill development for people with barriers to employment" [9]. The focus of this example is a Goodwill retail outlet. This facility represents a new model for Goodwill retail—with the drive-up donation center, sorting, selling, and baling for bulk sales all in one place.

This example also emphasizes the power of universal design in creating jobs for people with wide-ranging disabilities in that Goodwill's workforce is very diverse. The strategies implemented in this facility and the success of the diverse workforce employed at the facility bear witness to the efficacy of universal design.

Quoting from the Goodwill Industries of Southeastern Wisconsin website:

> One of the first things people recognize about Goodwill is the Goodwill Store—and rightly so, for the stores are key components of our history and our daily operations. Retail Operations also includes Secondary Markets—which process, recycle, and reuse donated goods not sold in the Goodwill stores. The revenue generated by Retail Operations provides funding for vocational training and services; training and employment opportunities are available throughout the division as well [10].

Figure 14.1 is a series of images showing various phases of the retail operation; the store itself, the drop-off, sorting, through to the retail display area and checkout.

The store is a modern building, looking like a regular discount store, located in West Bend, a suburb of Milwaukee, Wisconsin. West Bend is an affluent community and its donations are of good quality. The West Bend store has a fast turn-around. There are regular customers waiting every morning when the doors open, typically looking for specific items. People who shop at Goodwill stores have varied reasons. There are folks who buy and then resell at garage sales. Books are a popular resale item; they tend to move fast, and customers include many resellers, collectors, and fast readers who often take books and trade them for others at libraries.

The drop-off point for donations is a convenient drive-up location on the side of the building. A conveyor belt delivers the donated material to the first sorting station. At this station, items are sorted according to resale or bulk recycling. Resale items include books, jewelry, and clothes.

The largest volume of donated material is clothing. The first sort is into that suitable for resale and that bound for bulk baling and sale. In addition to employing the 5S strategy and workplace organization, Goodwill has instituted a collection of standardized work procedures. Visual controls are widely used throughout the facility to reinforce and support worker compliance. One image in Figure 14.1 shows one bulletin board with big block letters "KAIZEN." The bulletin boards are located at key workstations and provide, in words and diagrams, procedures relevant to that workstation. Color coding is used throughout the sorting process. One set of tags provides a timestamp for the clothes destined for resale. One image shows a chart on the wall with colored circles. These determine a weekly rotation of stock. The tags are labeled and color coded. After 3 or 4 weeks if the clothes are not sold, they are removed from the retail floor, baled, and sold as bulk. Note that 80 percent to 85 percent of merchandise sells within 1 week.

Clothes bound for resale are further sorted into children's and adult, male and female, according to size. A series of images shows the women's sorting station and a color-coding fixture used for size sorting. Once sorted, the clothes are hung on mobile racks, which are used to transport the clothes to the retail part of the store. The clothes are hung in an appropriate area, children's clothes, men and women's clothes, etc. The round circle tags inform employees as to the time an article has been in the store. As noted, clothes exceeding 3 to 4 weeks are removed from the retail side and baled for bulk sale.

When customers enter the store, they see a row of checkout counters and racks of merchandise. Signage throughout the store identifies product areas: books, jewelry, and of course, clothes. Not shown is the organization of the checkout stations. Each checkout station has a standard collection of tools and materials that are typically required for checkout: a pair of scissors, tape, marker, etc. The drawers have visual controls indicating where each item should be placed.

Chapter fourteen: Universal and accessible design in the workplace 195

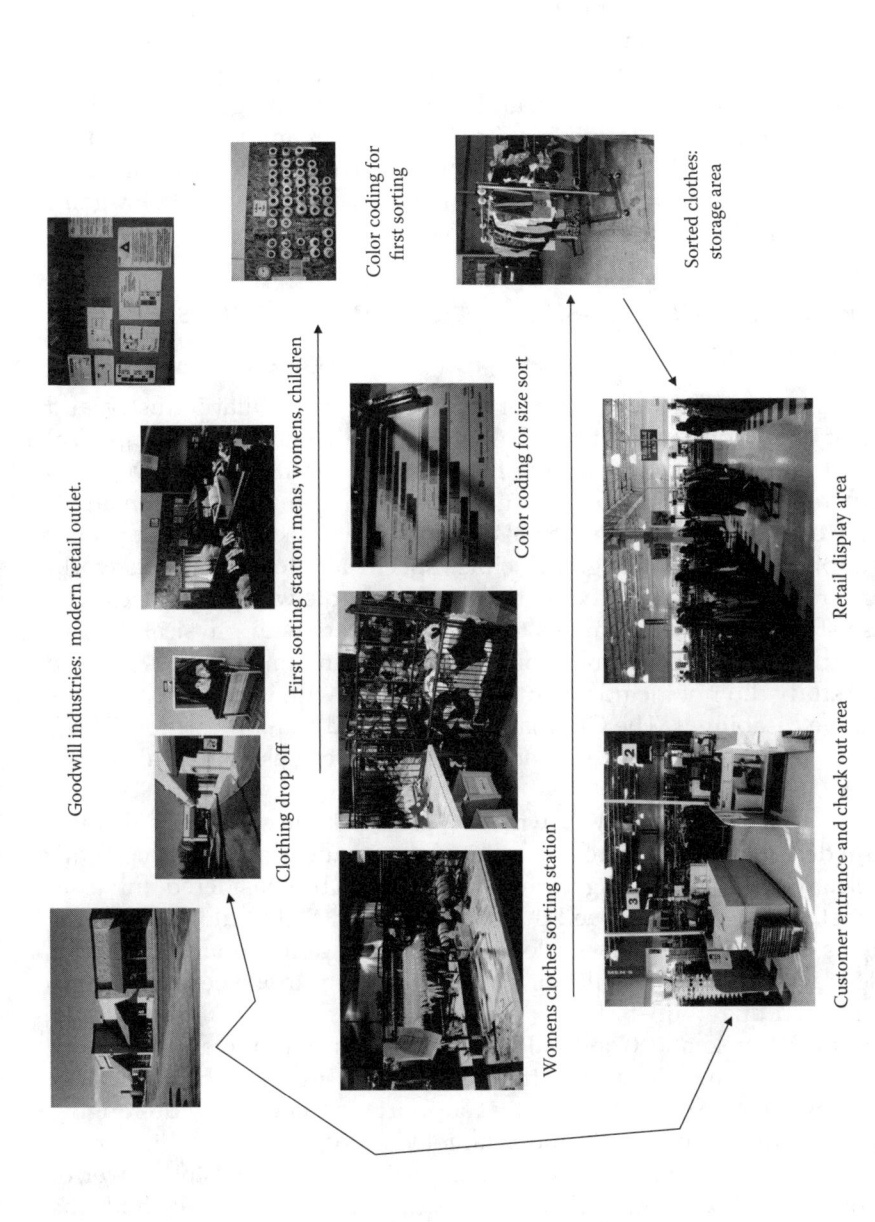

Figure 14.1 A Goodwill Industries retail outlet facility that exemplifies the use of universal and accessible design strategies to create jobs and improve the performance of all members of its diverse workforce.

As stated, Goodwill Industries of Southeastern Wisconsin, Inc., has embraced *kaizen* and quality control methods, and each of its operations is a model of *kaizen* approaches and successes. The methods are used at all locations, including the Great Lakes Naval Station–Food Service Operations, for which they have a contract to supply the food service. To sustain this level of operations, new employees receive *kaizen* training within 2 months. Retail has high employee turnover, and hence more training events are held for retail employees. This training is for all employees and is matched to the employee's cognitive abilities. Goodwill runs "Kaizen Events" at least yearly to review the past year's goals, the year's performance, and to set new goals for the upcoming year.

Nature of the industry, size of the company, and scale of operations

The nature of an industry, the size of a company within that industry, and its scale of operations all influence how a company might employ universal design in the workplace. Fuji Heavy Industries, Wal-Mart, and McDonald's are all very large multinational companies. Management at Fuji made a deliberate decision to apply universal design in the workplace. Management at Fuji applied universal design principles specifically to diversify its workforce and reduce barriers to the workplace for people with disabilities. The Fuji case study provides examples targeting workers with physical disabilities requiring a wheelchair and workers with hearing impairments [2]. The Fuji case study did not include mention of workers with visual impairments or cognitive disabilities. The Gunma Plant embraced lean production, TQM, and *kaizen* methods [2], but did not recognize these as universal design strategies.

Wal-Mart and McDonald's are more typical of large companies that use universal design methods, error-proofing, lean production, Six Sigma, TQM, *kaizen*, etc., without even being aware that these can be considered universal design methods. It can be argued that Wal-Mart and McDonald's used these techniques to encourage high employee turnover, suppress hourly wages for retail floor associates, while still being able to increase employee productivity with sustained quality [3–5, 11].

Art For A Cause and Goodwill Industries exemplify the application of universal design principles and methods in small businesses. The intent from the start was to create environments that support workers with wide-ranging abilities and skills. These applications targeted all types of disabilities, from physical to cognitive. Both applications explicitly recognized the power of lean production, Six Sigma, TQM, and *kaizen* methods to reduce both the ergonomic and cognitive demands of tasks and thereby render the workplace more accessible for all people while explicitly employing sound business practices to reduce cost, improve quality, and allow for the required productivity. In both cases, the production volume is relatively low but challenges the capabilities of the organizations given their funding, space, and staffing levels. The universal design methods are enabling them to be as productive as they can, given their financial and resource constraints.

Design tools for work

There are a variety of design tools available that can help designers deal with the complexity of today's workplace environment. For example, the Dassault Systemes collection of three-dimensional visualization tools from CATIA, DELMIA, and SIMULIA, including the human mannequins for ergonomic analysis, and their automation package that enables the three-dimensional simulation of complex manufacturing and assembly processes [1]. An article by Wanyama et al. [12] lists tools used and being developed for life-cycle engineering. While helping designers deal with design complexity, these tools do not explicitly contain design aids for accessibility or universal design.

Conclusions

Universal design is a natural design approach in our new global economy driven by e-commerce. Universal design is a cost-effective approach. Design, build, and sell products and services that as many people with as broad a spectrum of abilities can use without accommodation. For OXO,

> [T]he principles of universal design mean a salad spinner that can be used with one hand; liquid measuring cups that can be read from above without bending over; a toilet brush that bends to reach out-of-the-way places; a backlit oven thermometer that can be read easily through the window of an oven door; kettles with whistle lids that open automatically when tipped to pour; and tools with pressure-absorbing, non-slip handles that make them more efficient [13].

The International Association for Universal Design (IAUD) was started in Japan in 2002. Its membership includes not only Japanese companies, but also large multinational corporations. The IAUD seeks to expand the application of universal design principles for the betterment of humankind. The IAUD website makes the following statement:

> To design, from the start, equipment, a building, or living space so that it will be utilized by as many people as possible is called "universal design."
> In other words, universal design is a design that works for everyone, and this is how we should manufacture products all the time. Actually taking universal design into account will expand the customer base and improve customer satisfaction for a company, and, for the government, will be the foundation for

developing a city in cooperation with people from all walks of life. This is beneficial not only for the Japanese people but also for everyone in every part of the world [14].

Proponents of Six Sigma, *kaizen*, lean production, and other methods that fall under the accessible and universal design umbrella argue that as workers become more productive and fewer are needed to perform existing jobs, the opportunity exists for expansion and the creation of new jobs. This shifting of the labor force into new business opportunities does, in fact, occur under favorable economic conditions. OSHA requirements have improved worker ergonomic safety and reduced the ergonomic demands of jobs, rendering them more accessible. Error-proofing has eliminated safety risks as well as improved product and process quality. Accessible and universal design methods do, in fact, create jobs for and enhance the productivity of workers with disabilities, but as Erlandson as noted, this is not common knowledge within industry [15, 16].

Accessible and universal design principles are natural strategies for the design of work and service processes to support a global distribution of work and the associated incredibly diverse workforce. Design, manufacturing, assembly, and customer and service support functions can be distributed globally with a staggeringly diverse workforce. Transportable workplace and job designs that enable workers with different languages, different anthropometric features, and different skill levels to step into the process and with little or no training competently perform the required tasks are extremely desirable. Universal design in the workplace affords a design strategy aimed at creating transportable workplace designs that can effectively and efficiently serve the very diverse workforce found in a global economy.

Universal and accessible design strategies reduce both the physical and cognitive demands of tasks. Building knowledge into the environment, error-proofing, visual controls, reducing NVAAs, reducing process variability, all contribute to rendering jobs more accessible while ensuring sustained product and process quality. Because the employees are more productive, fewer are needed to do the same jobs. Employee training time is significantly reduced. The use of icons, color coding, and other visual controls reduces the need to read and simplifies other language requirements. Hence, the double-edged sword of universal and accessible design: the same principles and methods that can be used to create a more diverse workforce and create jobs for people with disabilities are also being used to sustain questionably ethical and often illegal employment practices for the least powerful and most vulnerable segments of our society [3–5].

References

1. Dassault Systemes, Products & Solutions, Web page showing their respective products, http://www.3ds.com/products-solutions/, 2006.

2. Y. Yano, Applying Universal Design Elements to the Workplace, presented at *Designing for the 21st Century III, An International Conference on Universal Design*, Rio de Janeiro, Brazil, 2004.
3. J. Dicker, *The United States of Wal-Mart*. New York: Jeremy P. Tarcher/Penguin, 2005.
4. E. Schlosser, *Fast Food Nation*. New York: Houghton Mifflin Co., 2004.
5. United Food and Commercial Workers International Union, a blog site critical of Wal-Mart, http://blog.wakeupwalmart.com/ufcw/2006/05/walmart_stores_1.html, 2006.
6. K. A. Krug, Artist's Business Becomes Tool to Help Disabled, in *The Oakland Press*. Pontiac, MI, 2006, pp. E1–E2.
7. A. Fluker, Tools for Good: Hand-Painted Products Put Art for a Cause on a Global Track, *Crain's Detroit Business*, 22, p. 29 and 33, October 23–29, 2006.
8. NSF, Enabling Technologies Laboratory: Student Design Projects, BES-0204099, 2002–2007.
9. Goodwill Industries of Southeastern Wisconsin, Inc. and Goodwill Industries of Metropolitan Chicago, Inc. This site is home to Goodwill Industries of Southeastern Wisconsin and Metropolitan Chicago, http://www.goodwillsew.com/, 2006.
10. Goodwill Industries of Southeastern Wisconsin, Inc. and Goodwill Industries of Metropolitan Chicago, Goodwill stores and donation centers, http://www.goodwillsew.com/page. asp?dbID=56&pageType=page, 2006.
11. Associated Press, Wal-Mart to Pay for Rest Break Violations, in *Detroit Free Press*. Detroit, MI, 2006, pp. 11A.
12. W. Wanyama, A. Ertas, H.-C. Zhang, and S. Ekwaro-Osire, Life-Cycle Engineering: Issues, Tools and Research, *Int. J. Computer Integrated Manufacturing*, 16, 307–316, 2003.
13. OXO, What We're About, http://www.oxo.com/about_what.php, 2004.
14. IAUD, International Association for Universal Design, http://www.iaud.net/en/member/index.html, 2004.
15. R. Erlandson, Business and Legal Conditions Supporting the Employment of Individuals with Disabilities, in *The Sourcebook of Rehabilitation and Mental Health Practice*, D. Moxley and J. Finch, Eds. New York: Plenum, 2003, pp. 51–60.
16. R. F. Erlandson, Accessible Design and Employment of People with Disabilities, in *The Sourcebook of Rehabilitation and Mental Health Practice*, D. Moxley and J. Finch, Eds. New York: Plenum, 2003, pp. 235–252.

chapter fifteen

The World Wide Web: Accessibility and universal design

Chapter goals:

- Present the World Wide Web Consortium (W3C) Web Accessibility Initiative (WAI) Guidelines.
- Relate the Web accessibility requirements of Section 508 of the Rehabilitation Act to accessible and universal design strategies and the WAI guidelines. Relate these guidelines to the principles of universal and accessible design.
- Show how the Internet and World Wide Web design communities have integrated universal and accessible design principles and strategies into their standard design procedures.
- Discuss emerging trends and implications for accessible and universal design.

Introduction

The Internet and World Wide Web (Web) design communities have integrated universal and accessible design principles and strategies into their standard design procedures. In this regard, these design communities exemplify the main goal of this book—that is, to mainstream universal and accessible design into all design practices and communities.

There are very pragmatic reasons why universal and accessible design have been so integrated and integral to Internet and Web design. These reasons include the Web's evolution from a communication and information novelty to part of the global communication and information infrastructure, the vision of the Web's founders with the Web a global and universally accessible entity, and the codification of this vision in the World Wide Web

Consortium's (W3C) goal of universal accessibility and the Web Accessibility Initiative (WAI) Guidelines.

Web accessibility is a very complex matter and Internet-based services are evolving at a rapid rate. This chapter only touches on some of the major issues and concerns. Accessible and universal design strategies targeting the Web exemplify the broader principles and strategies but are phrased and implemented for the particular product and services provided by the Internet or Web.

This chapter starts with a brief description the World Wide Web Consortium (W3C) and its Web Accessibility Initiative (WAI). The W3C is not the only standards and guideline setting and recommending group, but it has had a powerful influence on current laws, and its broad design principles are generally accepted by most developers. The WAI Content Accessibility Guidelines 1.0 are discussed and related to the universal design principles presented in previous chapters. Specific guideline checkpoints are presented as examples of universal and accessible design strategies targeting Web accessibility. The Section 508 accessibility requirements are presented and related to the WAI Content Accessibility Guidelines 1.0. The Section 508 requirements also are related to the universal design principles.

Emerging technologies such as grid services, the use of metadata, and the semantic Web are discussed in the context of universal and accessible design. A very significant development is that emerging Web design and authoring tools have built-in features that support and encourage the application of accessible and universal design strategies. A number of these new design and authoring tools are reviewed, along with Web accessibility checking tools and additional design resources.

Infrastructure

The Internet and World Wide Web (Web) have become part of the U.S. and global communication and information exchange infrastructure. As such, Web activity touches countless millions of lives everyday. Just as the telephony became part of the communications and information infrastructure and thereby essential to people lives, so too has the Internet and Web access. And just as a number of laws mandate the accessibility of telephone products and services for people with disabilities, laws and regulations are emerging mandating the accessibility of the Web.

In the United States, Section 508 of the Rehabilitation Act is the major legal driver for accessibility. While Section 508 specifically deals with federal websites, its influence is rapidly spreading. According to a 2003 survey by the Information Technology Technical Assistance and Training Center, of the Georgia Institute of Technology, 42 of the 50 states and the District of Columbia have adopted Section 508 (or very similar) standards [1]. The other states are expected to follow suit. Furthermore, universities and colleges across the country are implementing accessibility features, as are foundations, corporations, and most federally funded and supported websites.

The World Wide Web Consortium (W3C) and the Web Accessibility Initiative (WAI)

The World Wide Web is a dominant global entity and it presents a collection of unique design issues. The W3C develops specifications, guidelines, common protocols, software, and tools to promote the Web's evolution, ensure its interoperability, and lead it to its "full potential" [2]. The W3C is a forum for information, commerce, communication, and collective understanding, and has approximately 350 member organizations from around the world.

Tim Berners-Lee, inventor of the Web, founded the W3C in 1994. The W3C started at the Massachusetts Institute of Technology, Laboratory for Computer Science [MIT/LCS], with the collaboration of CERN (European Organization for Nuclear Research), and with support from DARPA (U.S. Defense Advanced Research Project Agency) and the European Commission. In April 1995, INRIA (Institute National de Recherche en Informatique et Automatique) became the first European W3C host, followed in 1996 by Keio University of Japan (Shonan Fujisawa Campus) in Asia. In 2003, ERCIM (European Research Consortium in Informatics and Mathematics) took over the role of the European W3C host from INRIA [2].

The W3C concentrates its efforts on three principle tasks. First, "the W3C promotes and develops its vision of the future of the World Wide Web" [2]. Second, the "W3C designs Web technologies to realize this vision, taking into account existing technologies as well as those of the future" [2]. Third, the W3C contributes to efforts to standardize Web technologies by producing specifications (called "Recommendations") that describe the building blocks of the Web. The W3C makes these Recommendations, and other technical reports, freely available to all [2]. These tasks are organized around three W3C long-term goals.

The first goal is universal access, that is, "to make the Web accessible to all by promoting technologies that take into account the vast differences in culture, languages, education, ability, material resources, access devices, and physical limitations of users on all continents" [2]. The second goal is to create a semantic Web, that is, "to develop a software environment that permits each user to make the best use of the resources available on the Web" [2]. The third goal is to develop a web of trust by guiding the "Web's development with careful consideration for the novel legal, commercial, and social issues raised by this technology" [2]. This discussion focuses primarily on the first long-term goal, that of universal access.

Design principles of the Web

Consumers tend to think of the Internet and the Web as the same entity. In fact, the Web is an application that is built on top of the Internet. Given this fact, the design principles of the Web are closely linked to those of the Internet. Three design principles emerge for Web design.

1. *Interoperability*: Specifications for the Web's languages and protocols must be compatible with one another and allow (any) hardware and software used to access the Web to work together.
2. *Evolution*: The Web must be able to accommodate future technologies. Design principles such as simplicity, modularity, and extensibility will increase the chances that the Web will work with emerging technologies such as mobile Web devices and digital television, as well as others to come.
3. *Decentralization*: Decentralization is without a doubt the newest principle and most difficult to apply. To allow the Web to "scale" to worldwide proportions while resisting errors and breakdowns, the architecture (like the Internet) must limit or eliminate dependencies on central registries [2].

These principles guide the work carried out within W3C activities.

Relationship of the Web content accessibility guidelines to accessible and universal design principles

The W3C has established the Web Accessibility Initiative (WAI) [3] to formally work toward universal accessibility. The WAI has developed and published accessibility guidelines [4], along with a collection of development tools, examples, and support material.

The Web content accessibility guidelines are intended to make the Web content more available to all users, regardless of their access mode, for example, desktop browser, voice browser, mobile phone, automobile-based personal computer, etc., or operating constraints such as noisy surroundings, under- or over-illuminated rooms, or in a hands-free environment. Web users also include people with physical or cognitive impairments. The following list is taken from the WAI website and notes that users [4]:

1. May not be able to see, hear, move, or may not be able to process some types of information easily or at all
2. May have difficulty reading or comprehending text
3. May not have or be able to use a keyboard or mouse
4. May have a text-only screen, a small screen, or a slow Internet connection
5. May not speak or understand fluently the language in which the document is written
6. May be in a situation where their eyes, ears, or hands are busy or interfered with (e.g., driving to work, working in a loud environment, etc.)
7. May have an early version of a browser, a different browser entirely, a voice browser, or a different operating system

All Web content designers and developers, page authors, and developers of authoring tools need to be aware of the Web content accessibility guidelines. There are 12 guidelines, and each guideline has a collection of checkpoints. Each guideline can be associated with one or more of the universal design principles. Table 15.1 shows the 12 guidelines and the associated universal design principles.

The goal of the W3C/WAI guidelines is universal accessibility to the Web. This goal can be restated in terms of the *equitable* universal design principle. That is, the Web should be equitable in that it should provide the same means of use for all users: identical whenever possible, equivalent when not possible. The Web and its products and services should avoid segregating or stigmatizing any user, making the design appealing to all users. Hence, all the WAI guidelines are aimed at Web equitability and the universal design principle *equitable* will be associated with each guideline.

Creating a stable and predictable Web environment and operation is another universal design principle that cuts across all the WAI guidelines. This follows because each of the guidelines reflects one of the universal design strategies associated with stability and predictability. Most fundamentally, the guidelines function to reduce the common cause variability of Web operations. Therefore, the universal design principle *stable and predictable* will also be associated with each guideline.

Table 15.2 provides a table of the checkpoints associated with Guideline 1. Each guideline has a unique collection of checkpoints. The respective checkpoints represent accessible design strategies for Web content design. Use of the checkpoint design strategy addresses the universal design principles associated with the guideline. Note that each checkpoint has an associated priority level. There are three levels of priority associated with different levels of accessibility. Table 15.3 presents the priority levels.

The WAI guidelines define three levels of conformance. Conformance level A means that all Priority 1 checkpoints are satisfied. Conformance level Double A (AA) means that all Priority 1 and Priority 2 checkpoints are satisfied. Conformance level Triple A (AAA) means that all Priority 1, 2, and 3 checkpoints are satisfied.

Figure 15.1 shows the Web Content Accessibility Guidelines (WCAG) Conformance Logos for levels A, Double A, and Triple A of the Web Content Accessibility Guidelines 1.0 [4]. A content provider claims Web content accessibility compliance with a conformance level by displaying the corresponding logo at the end of the Web page. Resources are available to help designers ensure compliance [5]. As discussed in the "Authoring tools" section in this chapter, there are a number of Web content accessibility checkers available.

Together, the WAI guidelines and associated checkpoints present a comprehensive collection of Web content accessibility design strategies. The conformance levels provide a measure of a website's accessibility. Use of the WAI guidelines and adherence to the conformance criteria are voluntary. However, Section 508 accessibility requirements, modeled on the WAI guidelines, are not voluntary for federal agencies and, as noted, for many state and local

Table 15.1 Summary of the WAI Guidelines [4] and Relationship to Universal Design Principles

WAI Guideline	Description	Universal Design Principle
		Common to all guidelines. Equitable, stable and predictable
Guideline 1. Provide equivalent alternatives to auditory and visual content.	Provide content that, when presented to the user, conveys essentially the same function or purpose as auditory or visual content.	Perceptible, flexible
Guideline 2. Do not rely on color alone.	Ensure that text and graphics are understandable when viewed without color.	Perceptible, flexible
Guideline 3. Use markup and style sheets and do so properly.	Mark up documents with the proper structural elements. Control presentation with style sheets rather than with presentation elements and attributes.	Cognitively simple
Guideline 4. Clarify natural language usage.	Use markup that facilitates pronunciation or interpretation of abbreviated or foreign text.	Cognitively simple
Guideline 5. Create tables that transform gracefully.	Ensure that tables have necessary markup to be transformed by accessible browsers and other user agents.	Flexible
Guideline 6. Ensure that pages featuring new technologies transform gracefully.	Ensure that pages are accessible even when newer technologies are not supported or are turned off.	Flexible
Guideline 7. Ensure user control of time-sensitive content changes.	Ensure that moving, blinking, scrolling, or auto-updating objects or pages may be paused or stopped.	Ergonomic, perceptible, flexible, error-proofing
Guideline 8. Ensure direct accessibility of embedded user interfaces.	Ensure that the user interface follows principles of accessible design: device-independent access to functionality, keyboard operability, self-voicing, etc.	Flexible
Guideline 9. Design for device independence.	Use features that enable activation of page elements via a variety of input devices.	Flexible

Table 15.1 Summary of the WAI Guidelines [4] and Relationship to Universal Design Principles (Continued)

WAI Guideline	Description	Universal Design Principle
Guideline 10. Use interim solutions.	Use interim accessibility solutions so that assistive technologies and older browsers will operate correctly.	Flexible
Guideline 11. Use W3C technologies and guidelines.	Use W3C technologies (according to specification) and follow accessibility guidelines. Where it is not possible to use a W3C technology, or doing so results in material that does not transform gracefully, provide an alternative version of the content that is accessible.	Equitable
Guideline 12. Provide context and orientation information.	Provide context and orientation information to help users understand complex pages or elements. Grouping elements and providing contextual information about the relationships between elements can be useful for all users. Complex relationships between parts of a page may be difficult for people with cognitive disabilities and people with visual disabilities to interpret.	Cognitively simple, memory, Hicks' law

Source: Web Content Accessibility Guidelines 1.0, W3C Recommendation 5. May–1999, Copyright © 1999, W3C (MIT, INRIA, Keio).

governmental units as well. In a very pragmatic way, Section 508 requirements are driving Web accessibility design activities in the United States.

Section 508 requirements

Subsection § 1194.22 (*Web-based intranet and internet information and applications*) of Section 508 is related to Web accessibility; Table 15.4 presents this subsection along with a note relating the subsection parts to the universal design principles. As with the W3C/WAI guidelines, the Section 508 requirements strive for equitable, and stable and predictable products and services; as such, the universal design principles *equitable* and *stable and predictable* are associated with each requirement.

The Access Board has explicitly stated the relationship between the requirements of §1194.22 and the Web Content Accessibility Guidelines 1.0.

Table 15.2 The Checkpoints Associated with Guideline 1 {provide equivalent alternatives to auditory and visual content} [4] along with their respective priority level.

Priority	Checkpoint and Description*
Priority 1	**1.1** Provide a text equivalent for every non-text element (e.g., via "alt," "longdesc," or in element content). *This includes* images, graphical representations of text (including symbols), image map regions, animations (e.g., animated GIFs), applets and programmatic objects, ASCII art, frames, scripts, images used as list bullets, spacers, graphical buttons, sounds (played with or without user interaction), stand-alone audio files, audio tracks of video, and video.
Priority 1	**1.2** Provide redundant text links for each active region of a server-side image map.
Priority 1	**1.3** Until user agents can automatically read aloud the text equivalent of a visual track, provide an auditory description of the important information of the visual track of a multimedia presentation.
Priority 1	**1.4** For any time-based multimedia presentation (e.g., a movie or animation), synchronize equivalent alternatives (e.g., captions or auditory descriptions of the visual track) with the presentation.
Priority 3	**1.5** Until user agents render text equivalents for client-side image map links, provide redundant text links for each active region of a client-side image map.

* Taken from [4], http://www.w3.org/TR/WAI-WEBCONTENT/ priorities, Web Content Accessibility Guidelines 1.0, W3C Recommendation 5 - May - 1999, Copyright © 1999 W3C (MIT, INRIA, Keio).

Table 15.3 The Priority Levels and Description

Priority	Description
Priority 1	A Web content developer must satisfy this checkpoint. Otherwise, one or more groups will find it impossible to access information in the document. Satisfying this checkpoint is a basic requirement for some groups to be able to use Web documents.
Priority 2	A Web content developer should satisfy this checkpoint. Otherwise, one or more groups will find it difficult to access information in the document. Satisfying this checkpoint will remove significant barriers to accessing Web documents.
Priority 3	A Web content developer may address this checkpoint. Otherwise, one or more groups will find it somewhat difficult to access information in the document. Satisfying this checkpoint will improve access to Web documents.

Source: From [4], http://www.w3.org/TR/WAI-WEBCONTENT/ priorities, Web Content Accessibility Guidelines 1.0, W3C Recommendation 5 - May - 1999, Copyright © 1999 W3C (MIT, INRIA, Keio).

Chapter fifteen: The World Wide Web: Accessibility and universal design

[W3C | WAI-A WCAG 1.0] Level A

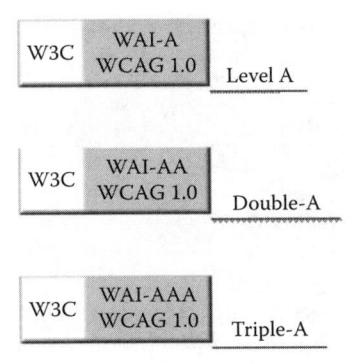

[W3C | WAI-AA WCAG 1.0] Double-A

[W3C | WAI-AAA WCAG 1.0] Triple-A

Figure 15.1 Shown are three Web Content Accessibility Guidelines (WCAG) Conformance Logos. These logos can be used by content providers to claim conformance to a specified conformance level of the Web Content Accessibility Guidelines 1.0. (*Source:* From http://www.w3.org/WAI/WCAG1-Conformance.html. Copyright © $Date: 2006/04/27 16:57:41 $, World Wide Web Consortium, Massachusetts Institute of Technology, European Research Consortium for Informatics and Mathematics, Keio University. All Rights Reserved. http://www.w3.org/Consortium/Legal/2002/copyright-documents-20021231.)

Note to §1194.22: 1. The Access Board interprets paragraphs (a) through (k) of this section as consistent with the following priority 1 Checkpoints of the Web Content Accessibility Guidelines 1.0 (WCAG 1.0) (May 5, 1999) published by the Web Accessibility Initiative of the World Wide Web Consortium:

Section 1194.22 Paragraph	WCAG 1.0 Checkpoint
(a)	1.1
(b)	1.4
(c)	2.1
(d)	6.1
(e)	1.2
(f)	9.1
(g)	5.1
(h)	5.2
(i)	12.1
(j)	7.1
(k)	11.4

2. Paragraphs (l), (m), (n), (o), and (p) of this section are different from WCAG 1.0. Web pages that conform to WCAG 1.0, level A (i.e., all priority 1 checkpoints) must also meet paragraphs (l), (m), (n), (o), and (p) of this section to comply with this section [6].

Table 15.4 § 1194.22 Web-Based Intranet and Internet Information and Applications

Requirement*	Universal Design Principle
	Common to all requirements. Equitable, stable and predictable
(a) A text equivalent for every non-text element shall be provided (e.g., via "alt," "longdesc," or in element content).	Perception Cognitive Flexible Efficient
(b) Equivalent alternatives for any multimedia presentation shall be synchronized with the presentation.	Cognitive Efficient Error-proof
(c) Web pages shall be designed so that all information conveyed with color is also available without color, for example, from context or markup.	Perception Cognitive Flexible
(d) Documents shall be organized so they are readable without requiring an associated style sheet.	Flexible Efficient Error-proof
(e) Redundant text links shall be provided for each active region of a server-side image map.	Flexible Efficient Error-proof
(f) Client-side image maps shall be provided instead of server-side image maps except where the regions cannot be defined with an available geometric shape.	Flexible Efficient
(g) Row and column headers shall be identified for data tables.	Efficient Error-proof
(h) Markup shall be used to associate data cells and header cells for data tables that have two or more logical levels of row or column headers.	Flexible Efficient Error-proof
(i) Frames shall be titled with text that facilitates frame identification and navigation.	Flexible Efficient Error-proof
(j) Pages shall be designed to avoid causing the screen to flicker with a frequency greater than 2 Hz and lower than 55 Hz.	Perception Health and safety
(k) A text-only page, with equivalent information or functionality, shall be provided to make a website comply with the provisions of this part, when compliance cannot be accomplished in any other way. The content of the text-only page shall be updated whenever the primary page changes.	Perception Cognitive Flexible Efficient Error-proof
(l) When pages utilize scripting languages to display content, or to create interface elements, the information provided by the script shall be identified with functional text that can be read by assistive technology.	Flexible Efficient Error-proof

Table 15.4 § 1194.22 Web-Based Intranet and Internet Information and Applications (Continued)

Requirement	Universal Design Principle
(m) When a Web page requires that an applet, plug-in, or other application be present on the client system to interpret page content, the page must provide a link to a plug-in or applet that complies with § 1194.21(a) through (l).	Flexible Efficient Error-proof
(n) When electronic forms are designed to be completed online, the form shall allow people using assistive technology to access the information, field elements, and functionality required for completion and submission of the form, including all directions and cues.	Flexible Efficient Error-proof
(o) A method shall be provided that permits users to skip repetitive navigation links.	Flexible Efficient
(p) When a timed response is required, the user shall be alerted and given sufficient time to indicate more time is required.	Perception Cognitive Flexible

* The requirement is taken from Federal Register, 36 CFR Part 1194, 2000 [6].

Together, the WAI guidelines and Section 508 software and Web accessibility requirements provide a collection of accessible and universal design strategies that address all the universal design principles. However, the WAI guidelines and Section 508 requirements are not the only source of standards, guidelines, or Web design recommendations.

Emerging products, standards, guidelines, and Web design recommendations

A variety of organizations have emerged and are promoting standards, guidelines, and recommendations with respect to Internet-based activities. This discussion is meant to be illustrative and is in no way exhaustive.

The European Commission's Information Society Technologies program (IST) conducted project Diffuse to provide neutral reporting on developments relating to standards and specifications [7]. The Diffuse project was funded "to provide a single, value-added, entry point to up-to-date reference and guidance information on available and emerging standards and specifications that facilitate the electronic exchange of information" [8].

The final conference of the Diffuse project focused on three Internet applications: the Web, grid services, and the semantic Web. "Web services connect computers and devices with each other using the Internet to exchange data and combine data in new ways" [7]. Grid services are evolving and vary, but the basic idea is that service creation and delivery occur through "coordinated resource sharing and problem solving in dynamic, multi-institutional virtual organizations" [7]. The semantic Web is a distributed machine, a computer system, "which should function so as to perform

Table 15.5 Selected Organizations that Promote Standards, Guidelines, and Recommendations for Internet-Based Activities

Organization	Current Activity
W3C/WAI	Web Accessibility Guidelines 2.0. These will address many of the concerns Web designers and developers have about the complexity, vagueness, and difficulties of the current Web Accessibility Guidelines 1.0 [28].
IETF – Internet Engineering Task Force	The IETF is engaged in the development of new Internet standard specifications. The IETF is not a corporation and has no board of directors, no members, and no dues. The IETF is not a traditional standards organization, although many specifications are produced that become standards. The IETF is made up of volunteers, many of whom meet three times a year to fulfill the IETF mission [29].
OASIS – Organization for the Advancement of Structured Information Standards	OASIS works on the development, convergence, and adoption of e-business standards [30]. OASIS was founded in 1993 and is a not-for-profit, international consortium with more than 3000 participants from over 600 organizations representing about 100 countries [30].
DCMI – Dublin Core Metadata Initiative	The DCMI is an organization dedicated to promoting the widespread adoption of interoperable metadata standards and developing specialized metadata vocabularies for describing resources that enable more intelligent information discovery systems [31].
OGSA – Open Grid Services Architecture	The Globus Project offers the Globus Toolkit, which provides designers with a collection of program development tools to help design OGSA-compliant grid structures [32].

socially useful tasks" [7]. Each of these applications has different design implications.

Table 15.5 provides a brief overview of some of the organizations addressed in the Diffuse project. This list is meant to be illustrative of the variety and scope of work to be done around the world. All of these organizations embrace the principles of universal access and accessibility.

The work of the DCMI (Dublin Core Metadata Initiative) with metadata, OASIS (Organization for the Advancement of Structured Information Standards) efforts with e-business, and the IETF (Internet Engineering Task Force) initiatives are all examples of early steps toward a "smarter" Web structure, one modeled upon the semantic Web concept. As such "smarter" Web structures evolve, it is critical that the new technologies do not create disabilities for large classes of users. Universal and accessible design strategies must be integrated into these technologies to reduce or eliminate environmentally generated disability. For this to occur, the design and authoring tools must support universal and accessible design principles and strategies. This important step is, in fact, happening.

Authoring tools

The term "authoring tool" refers to any software that can be used to produce content for publishing on the Web [9]. Authoring tools can enable and assist designers in the creation of accessible Web content through prompts, alerts, checking and repair functions, help files, and automated tools. The W3C *Authoring Tool Accessibility Guidelines 1.0* [9] list some examples of authoring tools. These include editing tools such as WYSIWYG (**w**hat **y**ou **s**ee **i**s **w**hat **y**ou **g**et) HTML and XML editors; word processors, desktop publishing programs, and other applications that allow the user to save material in a Web format; and programs that produce multimedia intended for use on the Web.

The W3C *Authoring Tool Accessibility Guidelines 1.0* were developed "to assist developers in designing authoring tools that produce accessible Web content and to assist developers in creating an accessible authoring interface" [9].

Traditional suppliers of software authoring tools are actively engaged in enhancing their capabilities with respect to accessible design tools and features. For example, Microsoft, in cooperation with HiSoftware, is offering owners of FrontPage a free licensed copy of HiSoftware's AccVerify SE™ [10]. AccVerify® SE™ for FrontPage provides verification and reports all errors and noncompliance with Section 508 and WAI conformance recommendations. Macromedia's recent version of its vector-based animation tool, Flash MX, provides designers with a collection of accessibility support tools and features [11]. Adobe's GoLive has a few accessibility features built into the software [12]. To enhance its accessibility design features, Adobe has partnered with SSB Technologies to provide a licensed copy of InSight LE for more accessibility checking. InSight LE will automatically identify Section 508 accessibility violations in Web pages created using GoLive [12].

An example authoring tool for the average consumer is the Accessible Web Publishing Wizard for Microsoft Office. It offers an alternative to the native Web publishing features in Microsoft Office for Word, PowerPoint, and Excel. It simplifies the task of converting PowerPoint presentations, Word documents, and Excel spreadsheets to accessible code that meets or exceeds Section 508 and W3C WCAG 1.0 Double-A requirements for accessibility by people with disabilities [13].

The CPB/WGBH National Center for Accessible Media (NCAM) has developed two tools for developers of Web- and CD-ROM-based multimedia for making their materials accessible to persons with disabilities—version 1.0 and 2.01 of the Media Access Generator (MAGpie)—for creating captions and audio descriptions for rich media [14]. The CPB/WGBH NCAM is a research and development facility dedicated to the issues of media and information technology for people with disabilities in their homes, schools, workplaces, and communities [14].

There are also a number of Web accessibility checking programs available. Bobby is a very popular Web checking system distributed by Watchfire Corporation. Bobby checks a website for Section 508 and WAI Version 1.0 guidelines compliance. Bobby provides a detailed report of noncompliance and potential noncompliance code [15]. The Watchfile Corporation website provides free online Web page checking [15].

UsableNet, Inc. [16], a Macromedia partner, created the free 508 Accessibility Suite, which extends Macromedia Dreamweaver's functionality to allow checking of Web documents' accessibility against proposed standards. UsableNet also offers LIFT and LIFTPro. LIFT works with Dreamweaver, providing a detailed analysis that makes up the free Section 508 analysis suite. With LIFT, one can select a set of accessibility rules one wants to test for—Section 508 or WAI. Using the analysis reports, a designer can go through the program code item-by-item, changing problem areas of the website at the code level of one's pages. UsableNet also has LIFT Online, which allows developers to test an unlimited number of live sites for an annual subscription fee.

United States government resources

The General Services Administration (GSA) (http://www.gsa.gov/) is charged with the dissemination of information regarding Section 508. The GSA manages a Section 508 website (http://www.section508.gov/) with the purpose of increasing understanding of the law and its implications, and providing strategies and resources for fulfilling the 508 standards. In addition, the website provides up-to-date information on events, training, and media coverage related to Section 508. The *Section 508 Buy Accessible Wizard* [17], provided by the Accessibility Forum, offers federal purchasing agents a tool to assess the accessibility of E&IT (electronic and information technology) products and services.

In addition to the Access Board's website [18], other federal agencies are providing accessibility and usability resource information. For example, the Department of Health and Human Services provides an extensive list of resources for Web authoring, checking, etc., at Usability.gov [19]. There is a widely dispersed collection of Section 508 checklists provided by various federal agencies that provide some assistance to designers of E&IT. Examples of such checklists can be found at the NASA website, which provides links to more than 40 federal agency websites dealing with Section 508 compliance [20]. The sites listed include the Department of Agriculture (http://www.usda.gov/), the Census Bureau (http://www.census.gov/), the Department of Defense (http://www.defenselink.mil/), the Department of Education (http://www.ed.gov/), the Federal Communications Commission (http://www.fcc.gov/cib/), and the Department Health and Human Services (http://www.hhs.gov/), just to name a few.

Additional resources

The W3C/WAI has a website, *Evaluation, Repair, and Transformation Tools for Web Content Accessibility* [21], that provides information and links to products and services that help Web designers design and author more accessible Web sites. The Accessible Design Resource Center (ADRC) website at Wayne State University has an extensive collection of accessibility resources [22]. The Special Needs Opportunity Windows (SNOW) project provides resources for Web authoring and checking [23], as does the HTML Writers Guild [24]. More resources are available at the University of Minnesota's *Web Accessibility Resources and Tools* website [25]. The Trace Research & Development Center, part of the College of Engineering, University of Wisconsin–Madison, is another rich source of information and material [26].

Conclusions

The W3C design philosophy and approach, as reflected in the W3C goals and WAI guidelines, exemplify the major goal of this book: to mainstream accessible and universal design principles and strategies. The W3C has as its first goal that of universal access. The W3C philosophy and goals have become part of the Internet and Web design culture. It is taken for granted that universal access is a fundamental design concern. Pragmatically, this means that universal design and accessible design are essential and fundamental components of the *thought process* associated with the design process discussed in Chapter 2.

An example that universal design is becoming part of the *thought process* of Web designers and programmers is that universal design has found its way into the curricula of programmers and Web designers. As part of the DO-IT (Disabilities, Opportunities, Internetworking, and Technology) project at the University of Washington, a collection of universal design Web page class projects have been developed and implemented [27]. This exemplifies the natural integration of universal design concepts into the education of programmers and Web designers, and provides another step in the mainstreaming process.

The creation of design and Web authoring tools that explicitly support and encourage the use of accessible and universal design strategies is a major force in promoting the practical application of these principles into actual designs. The Access Board is a repository for all accessibility-related laws, requirements, standards, training material, and resource material (http://www.access-board.gov/). However, while helpful, all the Access Board's checklists and resource materials do not constitute design tools as are being developed and implemented for Web design. As the internet and Web continue to grow and evolve it will present design challenges with respect to the goal of universal accessibility or equability. Having a collection of design and authoring tools that support accessible and universal design increases the likelihood of designing truly accessible products and services.

As the Web design community works to develop and ensure accessibility and universal access to the Web, a new global conceptual model of disability has been emerging, that is, the systems model as exemplified by the WHO-ICF model. Central to these new models is that disability can be reduced or eliminated as well as increased by environmental conditions. The Web design community recognizes the important role of the environment and the importance of universal Web access to the economic well-being of individuals and collectively to society.

Concurrent with the evolution of Web design are the various social and cultural movements that have emerged arguing strongly for not only the economic benefits of accessible and universal design, but also promoting an ethical imperative for universal design. The ethical imperative derives from the growing importance of E&IT products and services. The ethical imperative seeks to ensure that all people can fully participate in societal affairs. These arguments are moving the development and application of universal design principles from a legal and market-driven base back to a moral and ethical perspective.

Section IV of this book looks more closely at the moral and ethical imperatives that are further transforming accessible and universal design principles and the scope of their utilization.

References

1. ITTATC, Overview of State Accessibility Laws, Policies, Standards and Other Resources Available On-line, http://www.ittatc.org/laws/stateLawAtGlance.cfm, 2003.
2. W3C, About the World Wide Web Consortium (W3C), http://www.w3.org/Consortium/#mission, 2001.
3. WAI, Web Accessibility Initiative (WAI) — WAI Homepage, http://www.w3.org/WAI/, 2002.
4. WAI/W3C, Web Content Accessibility Guidelines 1.0, http://www.w3.org/TR/1999/WAI-WEBCONTENT-19990505/, 1999.
5. B. Kraus, *Web Content Accessibility Guidelines*, On-Line Course, Chap. 5, http://www.trynet.com/access/course/5.html. Try Net, 136 Pershing Ave., Nutley, NJ 07110, 2001.
6. Access Board, *Electronic and Information Technology Accessibility Standards*. Washington, D.C.: *Federal Register*, 36 CFR Part 1194, 2000.
7. Cover Pages, Diffuse Final Conference Focuses on Web Services, Grid Services, and Semantic Web, http://xml.coverpages.org/ni2002-10-29-b.html, 2002.
8. ICOM, The CIDOC Conceptual Reference Model, http://cidoc.ics.forth.gr/sites.html, 2004.
9. J. Treviranus, C. McCathieNevile, I. Jacobs, and J. Richards, Authoring Tool Accessibility Guidelines 1.0, 2004: W3C, http://www.w3.org/TR/WAI-AUTOOLS/, 2000.
10. HiSoftware, Microsoft and HiSoftware have formed a strategic relationship in order to bring you AccVerify® SE DS2 for FrontPage, http://www.hisoftware.com/msacc/, 2004.

11. C. Schmitt, Accessibility, Web Standards, and Authoring Tools, Vol. 22: *A List Apart Magazine*, http://www.alistapart.com/articles/tools/, 2002.
12. SSB Technologies, Adobe and SSB Technologies Bring Web Accessibility Solution to Adobe GoLive Users, http://askalice.ssbtechnologies.com/press/pr20010723.php, 2001.
13. Illinois Center for Instructional Technology Accessibility, Illinois Accessible Web Publishing Wizard for Microsoft® Office, http://cita.disability.uiuc.edu/software/office/, 2003.
14. NCAM, Media Access Generator (MAGpie), http://ncam.wgbh.org/webaccess/magpie/, 2004.
15. Watchfire Corporation, Welcome to the Bobby Online Free Portal, http://bobby.watchfire.com/bobby/html/en/index.jsp, 2004.
16. UsableNet. Inc, UsableNet Home Page, http://www.usablenet.com/, 2004.
17. Accessibility Forum, Buy Accessible Wizard, http://www.buyaccessible.gov/, 2004.
18. Access Board, "Access Board. This is the homepage URL for the Access Board. The Access Board is a key resource for accessible design issues," http//www.access-board.gov/, 2001.
19. Usability.gov, accessibility resources, http://www.usability.gov/accessibility/, 2004.
20. NASA, Links-Government, http://section508.nasa.gov/resources/resources_link_g.htm, 2002.
21. WAI, Evaluation, Repair, and Transformation Tools for Web Content Accessibility, http://www.w3.org/WAI/ER/existingtools.html, 2004.
22. R. Erlandson, ADRC-Home Page, http://www.etl-lab.eng.wayne.edu/adrc/, 2005.
23. SNOW, Special Needs Opportunity Windows, Evaluation & Repair Tools, http://snow.utoronto.ca/access/tools/, 2004.
24. HTML Writers Guild, Inc., The Accessible Web Author's Toolkit, http://aware.hwg.org/tools/, 1999.
25. Computer Accommodations Program — University of Minnesota, Web Accessibility Resources and Tools, http://cap.umn.edu/ait/Web/ResourcesAndTools.html, 2001.
26. Trace R&D Center, http://trace.wisc.edu/about/, 2005.
27. S. Burgstahler, Universal Design of Web Pages in Class Projects, http://ww.washington.edu/doit/Brochures/Technology/universal_class.html, 2005.
28. B. Caldwell, W. Chisholm, G. Vanderheiden, and J. White, Web Content Accessibility Guidelines 2.0, W3C Working Draft, http://www.w3.org/WAI/EO/Drafts/standard-harmon.html, 2003.
29. S. E. Harris, The Tao of IETF: A Novice's Guide to the Internet Engineering Task Force, http://www.ietf.org/tao.html, 2001.
30. OASIS, OASIS Home Page, http://www.oasis-open.org/home/index.php, 2004.
31. DCMI, The Dublin Core Metadata Initiative, http://dublincore.org/, 2004.
32. Globus Alliance, Towards Open Grid Services Architecture, http://www.globus.org/ogsa/, 2004.

section four

Ethical considerations and conclusions

Section goals:

- Explore the relationship between universal design and the ethics of disability.
- Present three definitions of universal design and show how these definitions represent an evolution of universal design toward a more systems theoretic framework.
- Explore the relationship between universal design and an individual's ability to fully partake and participate in the activities of society.
- Discuss the changing role of designers with respect to the evolving definition of universal design.
- Argue for the mainstreaming of universal design based on the perspective that advocates the full participation of all people in the affairs of society.

Scope of the discussion

In the United States, a common rationale for employing accessible and universal design is that "it's the right thing to do!" This is an ethical argument and a very weak one at that. It is weak in that it is too vague to be useful. Chapter 16 of this section argues that if an individual's life is deemed "of value" by society and if we live in a society that advocates the full participation of all its citizens in the activities and affairs of that society, then following the systems model of disability, society must provide the infrastructure and support necessary for such full participation to be possible.

The focus in Sections II and III was on the use of accessible and universal design principles on the "Products and Technology" domain of the WHO-ICF framework (that is, the traditional targets for design activities). Chapter 16 takes this relational chain one step further and considers the ethical impact on a person of the environment. It is argued that our environment, in the broadest sense and consistent with the WHO-ICF framework, plays a significant role in defining an individual's value to society. That is, the environment plays a significant role, from an ethical perspective, in defining a person's societal value.

Chapter 17, the concluding chapter of the book, brings the discussion back to where it began, but with a difference, in that we now explicitly consider economic, societal, and political influences on the practice of design. The practice of design takes place within a highly interconnected and complex global system, and this context is changing the role of the designer as well as pressuring universal design to assume a broader systems theoretic framework. The pressures to move universal design toward a systems framework and also toward a more inclusive integration into the design community derive from not only economic, social, and political pressures, but also from a new political ethics that seeks to ensure the full potential of all people within the affairs of society. Based on these arguments, the ethical imperative for mainstreaming accessible and universal design shifts from "it's the right thing to do" to "it's the necessary thing for a society to do so as to ensure the full participation of all its citizens in the activities and affairs of that society."

While Chapter 17 provides a summary and conclusion of the book's major points, it does so within an expanding economic, social, political, and ethical context. In so doing, it is the author's desire to challenge the readers' (1) conceptualization of the practice of design, and (2) self-concept—that is, your role as a designer.

chapter sixteen

Ethics and universal design

Chapter goals:

- Reexamine the ethics of disability from a personal perspective to a broader societal perspective of citizens being able to fully participate in society's activities and affairs.
- Show how the environment can influence a person's societal value.
- Present three definitions of universal design showing an evolution and expansion of the concept that conforms to the broader societal ethics of disability.

Ethics

Ethics are a hot topic. The Securities and Exchange Commission has been very busy investigating corporate conduct and imposing record fines and penalties [1]. The Justice Department is equally busy dealing with corporate fraud [1]. The *New York Times* attracted international attention for plagiarism by one of its most respected journalists [2]. Then there was the national attention and ethical debate generated by the removal of Terri Schiavo's feeding tube [3]. These stories highlight the meaning of ethics: "The discipline dealing with what is good and bad or right and wrong or with moral duty and obligation" [4].

Ethics of disability

> Singer has stated that he insists he doesn't want to kill me. He simply thinks it would have been better, all things considered, to have given my parents the option of killing the baby I once was, and to let other parents kill similar babies as they come along and thereby avoid the suffering that comes with lives like mine and satisfy the reasonable preferences of parents for a different kind of child. It has nothing to do with me. I should not feel threatened [5].

So says Harriet McBryde Johnson. Johnson received considerable national attention after the publication of a debate between Johnson and Princeton University's controversial bioethicist Peter Singer in *The New York Times Magazine* [5].

The objective of this section is to place the ethics of disability in a human context. McBryde Johnson is an articulate spokesperson, not only for herself, but for all people with impairments and consequently disabilities. The material in this section is drawn primarily from two sources, an April 12, 2003, article by Jennifer Berry Hawes in *The Post and Courier* [6], and the February 16, 2003, article in *The New York Times Magazine* [5].

McBryde Johnson is an attorney. She runs a solo law practice in Charleston, South Carolina, which mostly handles benefits and civil rights claims for poor and working people with disabilities [6]. From early childhood she has dealt with a progressive, congenital, neuromuscular disease. McBryde Johnson's generation was among the first to survive to, as she puts it, *"such decrepitude,"* because the availability of new antibiotics allowed individuals with weakened respiratory systems to survive childhood pneumonias [5].

McBryde Johnson is often the recipient of comments such as, "If I had to live like you, I think I'd kill myself" [5]. To McBryde Johnson, the word "overcome" has an annoying quality. "It's annoying because it conveys an attitude by nondisabled people whose pity makes them applaud a disabled person who so much as gets out of bed. ... It hits me real often from strangers and even people who know me pretty well because the expectations are so low that they insist on being amazed" [6].

McBryde Johnson's response to such comments and attitudes is: "I used to try to explain that in fact I enjoy my life, that it's a great sensual pleasure to zoom by power chair on these delicious muggy streets, that I have no more reason to kill myself than most people. But it gets tedious. God didn't put me on this street to provide disability awareness training to the likes of them" [5]. She goes on to say that even when she tried to talk to such people, to raise awareness, she felt that "they don't want to know. They think they know everything there is to know, just by looking at me. That's how stereotypes work. They don't know that they're confused, that they're really expressing the discombobulation that comes in my wake" [5].

People look at McBryde Johnson, form impressions, and jump to conclusions comparing their lives to hers. The debate between Professor Singer and McBryde Johnson generalizes from this observation/impression-forming process to the broader questions of the value and quality of one's life. The debate took place at Princeton University. The earlier quotation summarizes Professor Singer's basic position. Quoting McBryde Johnson:

> He wants to legalize the killing of certain babies who might come to be like me if allowed to live. He also says he believes that it should be lawful under some circumstances to kill, at any age, individuals with cognitive impairments so severe that he doesn't consider them "persons." What does it take to be a person?

Awareness of your own existence in time. The capacity to harbor preferences as to the future, including the preference for continuing to live.

At this stage of my life, he says, I am a person. However, as an infant, I wasn't. I, like all humans, was born without self-awareness. And eventually, assuming my brain finally gets so fried that I fall into that wonderland where self and other and present and past and future blur into one boundless, formless all or nothing, then I'll lose my personhood and therefore my right to life. Then, he says, my family and doctors might put me out of my misery, or out of my bliss or oblivion, and no one count it murder.

Speaking for people with disabilities McBryde Johnson goes on to ask: Are we "worse off"? I don't think so. Not in any meaningful sense. There are too many variables. For those of us with congenital conditions, disability shapes all we are. Those disabled later in life adapt. We take constraints that no one would choose and build rich and satisfying lives within them. We enjoy pleasures other people enjoy, and pleasures peculiarly our own. We have something the world needs. {*inserted by author} [5]

In *The Future of the Disabled in Liberal Society: An Ethical Analysis*, Reinders considers these same issues—life, death, and human value [7]. Reinders' book grew out of a 1996 request by the Dutch Association of Bioethics to write an essay on the ethical implications of human genetics and implications for people with mental disability. The question he was asked to address: should we prevent disabled lives? Genetic testing allows us to currently identify which couples might be at risk of parenting children with disabilities, and whether or not a fetus has a genetic or developmental disability. Societal attitudes and values, as investigated and articulated by Reinders, and the laws such attitudes and values spawn, are part of the broader environment within which we all exist.

Reinders notes that the "medical paradigm that has dominated the bioethical agenda not only renders the existence of disabled people intrinsically problematic, it turns to 'preventions' as the obvious solution to the problem" [7]. The medical paradigm, with its focus on prevention, gives rise to the following questions. Given the existence of genetic testing and the technical ability to identify potentially disabled people, should we not prevent their conception or birth so as to not inflict undue suffering and hardships not only on the disabled individual, but also their family, friends, and society? This line of argument also raises issues for the surviving disabled. How much value do you place on the life of a person who should not be here? As seen in McBryde Johnson's comments, these questions are not just hypothetical, but are being addressed and played out everyday by people around the world.

Reinders' observations and conclusions agree with and support McBryde Johnson's comments about the mixed messages she has received. These mixed messages include the following. You are here and as such we value and respect you as a person. You have rights and liberties as afforded all citizens. On the other hand, everyone would have been better off if you were not here at all. Reinders argues that society today finds itself faced with these two conflicting views of disability [7].

Given that McBryde Johnson and other people with disabilities are living and part of our population mix, the environment works in yet another way to assign worth to people. For example, McBryde Johnson uses a powered wheelchair for mobility and computer technology for written communications. These and other environmental entities (technologies) enable her to function effectively as an attorney and advocate for the disability community. It can be argued that her native endowments, such as intelligence and fortitude, in combination with a variety of environmental conditions, raise her societal value. In fact, this argument holds for all people, not just people with disabilities.

Native endowments, the environment, and disability

Section III presented universal design principles and strategies with the premise, supported by examples, that their utilization creates more accessible and less disabling environments. In Sections I, II, and III, it was argued that broader use of accessible and universal design would lead to improved functional capabilities for all people. The emphasis was on enhancing and improving human functional capabilities—both physical and cognitive. This chapter shifts the focus of attention from the impact of the environment on human functional capabilities to the impact of the environment on the perceived value of a person's native endowments, that is, it considers environmental impact from an ethical perspective.

Human beings are born with native endowments of various kinds: native intelligence, natural abilities, behavioral dispositions, etc. Native endowments are potentials, promises of what can be realized in a given environment, that is, social, political, and technological conditions. When realized, an individual's potential can take but one or a limited number of a larger number of possibilities. Education and training opportunities represent a small selection of the possibilities that might have been fulfilled. John Rawls, considered by many as the most important political philosopher of the second half of the 20th century, notes that the realization of potentials, created by one's natural endowments, are directly related to the "social attitudes of encouragement and support, and institutions concerned with their early discipline and use" [8]. Rawls further argues that conceptions of ourselves, our aims and ambitions, and ultimately our realized abilities and talents, derive from our personal history, opportunities, social position, and the influence of good or bad luck [8].

Chapter sixteen: Ethics and universal design

Technology contributes to one's environment, but the nature, capabilities, and targeted use of the technology are influenced by the broader social and political components of one's environment. However, technology also influences social and political thinking and values. For example, it is now commonly accepted that a blind person, with the appropriate technology such as screen readers, text-to-Braille systems, etc., can perform clerical and business tasks as well as a sighted person. This expectation evolved and became commonly accepted as more and more people experienced, either by using or observing its use, the effects of the technology. In a very real and pragmatic way, the perceived value of a blind person as an office employee increased as technology rendered him less disabled.

Rosenberg, in his paper entitled "The Political Philosophy of Biological Endowments: Some Considerations," presents an interesting example of how the environment can influence the value placed on a person's native endowments [9].

> In a population of Polynesians with no interest in coconuts, an agile tree climber will have no capital advantage. And in one in which all crave coconuts and can climb equally well, he will have no advantage either, though in both cases his biological endowment remains unaltered. ... But things might be quite different. Suppose the coconut trees are too high for any method but climbing, and there is nothing to eat but coconuts. And suppose that time preference is quite high because, lacking coconuts today, the population will starve to death tomorrow. And suppose, finally, that everyone has acrophobia and only a handful have a talent for climbing. Here, the present value of the talent becomes extremely high [9].

Rosenberg concludes that the environment plays a significant role in society's identification of a native endowment as significant, beneficial, benign, or harmful. Recall that the World Health Organization's systems conceptual model of disability holds that environmental conditions can increase or decrease one's disability with respect to activities within that environment. If the environment renders a person less disabled with respect to a given activity, then it is reasonable to conclude that the person's perceived value, with respect to the activity, will increase. Figure 16.1 illustrates these relationships diagrammatically.

Moulton et al., in *Accessible Technology in Today's Business: Case Studies for Success* [10], present examples from a variety of businesses where assistive and enabling technology created jobs for or greatly improved the performance of people with disabilities. These case studies demonstrate how technology can, in fact, increase a person's social value within a business environment.

If we subscribe to the World Health Organization's conceptual model of disability and we agree that universal design is a way to decrease or eliminate

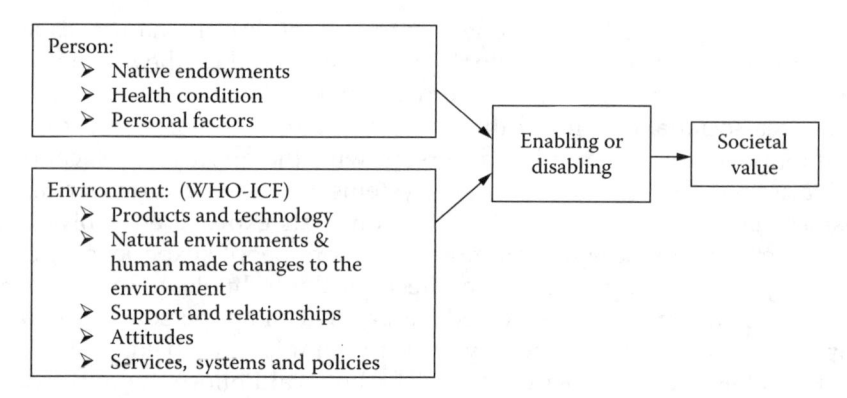

```
┌─────────────────────────────┐
│ Person:                     │
│   ➢ Native endowments       │
│   ➢ Health condition        │──┐
│   ➢ Personal factors        │  │   ┌──────────┐   ┌──────────┐
└─────────────────────────────┘  │──▶│ Enabling │──▶│ Societal │
                                 │   │    or    │   │  value   │
┌─────────────────────────────┐  │   │ disabling│   │          │
│ Environment: (WHO-ICF)      │  │   └──────────┘   └──────────┘
│   ➢ Products and technology │──┘
│   ➢ Natural environments &  │
│     human made changes to   │
│     the environment         │
│   ➢ Support and relationships│
│   ➢ Attitudes               │
│   ➢ Services, systems and   │
│     policies                │
└─────────────────────────────┘
```

Figure 16.1 Conceptual model derived from the WHO-ICF relating person–environment relationships to functional change (enabling or disabling) and a corresponding change of the societal value of the person.

environmentally induced disability, then is there not an ethical imperative for designers and planners to utilize universal design principles and strategies? The answer to this question must be yes if we accept the arguments presented in the previous paragraph. Not only is an individual with a disability eligible for all the rights and privileges associated with personhood, but additionally, the designed or technological elements of the environment can enhance human functioning and thereby increase the individual's perceived value, while concurrently raising societal expectations and establishing new societal norms. This is a powerful conclusion.

Universal design: Expanding the scope

The foregoing presented an argument for an ethical imperative that designers and planners utilize universal design principles and strategies. Acceptance of this argument would signify a shift in societal thinking, that is, from the need for accessible design prescribed and mandated by law, to the mainstreaming of universal design because it is not only good design, but ethically required. Evidence will now be presented that this transformation of societal thinking is already taking place.

Three definitions of universal design will be provided wherein each additional definition broadens the scope of universal design. The first definition, provided earlier and used in the preceding chapters, is derived from universal design's relatively narrow architectural roots. The definition focuses on the built environment and designed entities, and arose concurrently with the social model of disability that viewed disability as an environmental condition. In this context; *universal design can be defined as the design of entities that can be used and experienced by people of all abilities, to the greatest extent possible, without adaptations* [11].

The second definition of universal design is provided in a Council of Europe resolution proposing the introduction of universal design principles into the curricula of all occupations working on the built environment {Resolution ResAP(2001)1} [12]. The aim of the resolution is to guarantee utilization of universal design so as "to ensure equal chances of participation in economic, social, cultural, leisure and recreational activities, everyone of whatever age, size, and ability must be able to access, use and understand any part of the environment as independently and as equal to others as possible" [12].

The Council of Europe, Committee of Ministers, provides the following definition of universal design. "Universal design is a strategy, which aims to make the design and composition of different environments and products accessible and understandable to as well as usable by everyone, to the greatest extent in the most independent and natural manner possible, without the need for adaptation or specialized design solutions" [12]. This definition is significant because it moves the notion of universal design into a more political context.

This definition is a significant expansion of the first, in that it views universal design from a much broader societal context and sees universal design as a way to bring about a fuller "participation in economic, social, cultural, leisure and recreational activities" for all people. Further, this definition explicitly uses the terms environments, understandable, and usable, and as such expands the scope of universal design from the first definition. Note that the universal design principles and strategies presented in Section III also conform to this definition of universal design.

The integration of universal design principles into the curricula of designers and planners would place universal design in the mainstream of European design and establishes universal design as good design, as the expected design, as the norm. The European Concept for Accessibility Network makes the case for good design. In discussing diversity in the years 2000 and beyond, they urge Europeans to "no longer talk about the specific needs of certain categories of people, but talk about human functioning. We should look at every aspect of human functioning, without categorizing. ... Accessibility will lose its stigma and become a mainstream issue. We won't need terms like Design for All or Universal Design anymore. We will only refer to good design and bad design" [13].

The mainstreaming of universal design by the World Wide Web (Web) design community also gives evidence to the shift in societal thinking and is consistent with the European Council's rationale for and definition of universal design. As presented in Chapter 15 of Section III, universal access is a fundamental goal of the W3C Web Accessibility Initiative (WAI). Tables 15.1 and 15.4 demonstrate that the WAI guidelines and Section 508 guidelines incorporate all universal design principles. Emerging Web design and authoring programs contain specific accessible design tools and WAI and Section 508 compliance checking. Hence, the explicit goal of universal Web accessibility by the W3C and the explosion of Web authoring and design tools that support and encourage universal design exemplifies the integration of universal design into mainstream thinking about the Web.

Corporations are also recognizing the links between the environment and disability, and are taking very concrete and pragmatic steps to address the issues. Toyota has embraced the utilization of universal design across all its products and services [14]. This includes cars and transportation as well as housing [15]. Toyota has built the TOYOTA Universal Design Showcase in downtown Tokyo to explain and promote universal design [14]. OXO International, a renowned and widely recognized maker of kitchen and culinary utensils, uses universal design principles throughout its entire product line and is a strong advocate of universal design [16].

The following quote from the preface to Southwestern Bell Communications (SBC) ethical code is remarkably similar to the vision of the European Council Resolution.

> Universal design is critical to millions of Americans who depend on telecommunications accessibility for employment, education, social interaction, recreation and other life activities. Access to telecommunications allows persons with disabilities greater participation in society, without which they face pervasive challenges. SBC's commitment to universal design principles is a tangible demonstration of the value SBC places on the worth and dignity of all individuals, including people with disabilities. SBC is committed to universal design [17].

This statement explicitly recognizes the importance of telecommunication products and services in modern life and that lack of telecommunications accessibility can be disabling and limit a person's ability to participate fully in society. This view is consistent with the social and systems conceptual model of disability in that the environment can create disabilities or reduce disabilities. From the perspective of these models, universal design is the preferred design approach. Universal design has at its core an ethic of inclusiveness. This ethic of inclusiveness recognizes human differences and seeks to create entities that support a diversity of human capabilities. It is from this ethic of inclusiveness that the third definition of universal arises.

The third, and final, definition of universal design is a significant generalization and as such is a more thought-provoking conceptualization. Weisman in the paper entitled "Creating Justice, Sustaining Life: The Role of Universal Design in the 21st Century" offers the following definition of universal design along with a question. "If we were to define universal design as a values-based framework for design decision making, based on an ethic of inclusiveness and interdependence that values and celebrates human difference and acknowledges humanity's debt to the earth, how might this effect the design and use of the built environment" [18]? This is the most general and far-reaching definition of universal design. It is a definition that moves universal design squarely into the political arena. Weisman argues that, "While the focus on accessibility that gave rise to the universal design movement has begun to produce significant innovation in product and graphic design ... it is time to move beyond the letter of the law to the spirit of the law; to shift our focus from redressing human and environmental problems through remedial design to preventing problems through holistic design" [18].

Conclusions

Society's conceptualization of universal design is evolving; in particular, it is expanding from a relatively narrow scope to definitions that include broader ethical values and social consequences. The accessible and universal design principles presented in Sections II and III describe how designed entities can enhance and improve human functioning. This chapter argues that the environment places a social value on people. If technology enables a person to function at a higher, more proficient level, the person's value to society is seen to increase. These conceptualizations of universal design envision outcomes in which all people, including people with disabilities, can enjoy full participation in community life.

Societal values, in particular as reflected in our laws, strive to create an environment that supports fuller participation for all citizens. However, McBryde Johnson's personal experiences bear witness to the fact that people with disabilities generally do not enjoy full participation in community life. Accessible design, as driven by legal mandates, guidelines, and regulations, does increase accessibility in the regulated contexts. Universal design has the potential to greatly increase everyone's involvement and participation in community life. The next chapter revisits the societal linkages that influence the practice of design. It specifically looks at the interplay between designed entities, social values, political expressions, and evolving conceptualizations of universal design.

References

1. MacNeil and Lehrer, "Corporate Ethics: An Online NewsHour Special Report," Copyright © 2003 MacNeil/Lehrer Productions: http://www.pbs.org/newshour/bb/business/ethics/, 2003.
2. T. Kunkel, "Dean Thomas Kunkel Statement on New York Times Plagiarism Case," Copyright © 2003 Philip Merrill College of Journalism: http://www.journalism.umd.edu/newrel/03newrel/nytimesblair03.html, 2003.
3. M. Roig-Franzia, "Justices Decline Schiavo Case, Options Dwindle for Those Trying to Keep Florida Woman Alive," in *Washington Post*. Washington, D.C., 2005, pp. A01.
4. Merriam Webster, *Webster's Third New International Dictionary of the English Language Unabridged*. Chicago, IL: William Benton, 1966.
5. H. McBryde Johnson, "Unspeakable Conversations," in *The New York Times Magazine*: http://community.webtv.net/stigmanet, 2003.
6. J. Berry Hawes, "Harriet McBryde Johnson," in *The Post and Courier*. Charleston, 2003.

7. H. S. Reinders, *The Future of the Disabled in Liberal Society*. Notre Dame, IN: University of Notre Dame Press, 2002.
8. J. Rawls, *Justice as Fairness*. Cambridge, MA: The Belknap Press of Harvard University Press, 2001.
9. A. Rosenberg, "The Political Philosophy of Biological Endowments: Some Considerations," in *Equality of Opportunity*, Ellen F. Paul et al., Eds. Oxford: Basil Blackwell, 1987, pp. 1–31.
10. G. Moulton, L. Huyler, J. Hertz, and M. Levenson, "Accessible Technology in Today's Business: Case Studies for Success," Redmond, WA: Microsoft Press, 2003.
11. Center for Accessible Housing, "Accessible Environments: Toward Universal Design," Raleigh, NC: North Carolina State University, 1995.
12. Council of Europe Committee of Ministers, "Resolution ResAP(2001)1 on the introduction of the principles of universal design into the curricula of all occupations working on the built environment. Adopted by the Committee of Ministers on 15 February 2001, at the 742nd meeting of the Ministers Deputies, http://www.fortec.tuwien.ac.at/bk/BK-DOK-UnivDes.htm, 2001.
13. European Concept for Accessibility Network, Web home page: European Concept for Accessibility Network, Text Sites: History (http://www.eca.lu/history.html); and Definitions (http://www.eca.lu/def.htm), http://www.eca.lu/, 2001.
14. Toyota, "TOYOTA Universal Design Showcase," http://www.megaweb.gr.jp/Uds/English/guide.html, 2004.
15. Toyota, "Toyota's initiatives in universal design," http://www.toyota.co.jp/en/environmental_rep/04/download/pdf/p67.pdf, 2004.
16. College of Design, "OXO International Becomes a Universal Design Icon; Case Study," NC State University, College of Design: http://design.ncsu.edu/cud/proj_services/projects/case_studies/oxo.htm, 2000.
17. SBC, SBC: *Universal Design Policy*, Southwestern Bell Communications, 1998.
18. L. K. Weisman, "Creating Justice, Sustaining Life: The Role of Universal Design in the 21st Century," presented at Adaptive Environments Center 20th Anniversary Celebration, The Computer Museum, Boston, April 10, 1999, http://www.adaptenv.org/examples/article2.php?f=4.

chapter seventeen

Universal and accessible design from a social and political perspective

Chapter goals:

- Reexamine the linkages between design and broader societal concerns.
- Examine universal and accessible design from a social and political perspective.
- Examine the role of the designer.
- Argue for the mainstreaming of universal design from an ethical perspective that advocates the full participation of all its citizens in the activities and affairs of society.

Linkages

An old proverb goes something like this:

> As a child I played in a tree close by my home. I moved away.
> As an adult I came back home and saw my play tree. It was the same, but different.

We started this book with a discussion of how the practice of design does not take place in a vacuum, but rather is part of a rich tapestry of interdependence among technology, laws, ethical values, and society's conceptualization of disability. We end this book with a continuation of this same theme, but "with a difference."

Figure 17.1 starts with the elements of Figure 1.1 and adds economic considerations, and societal and political pressures to the mix. The design

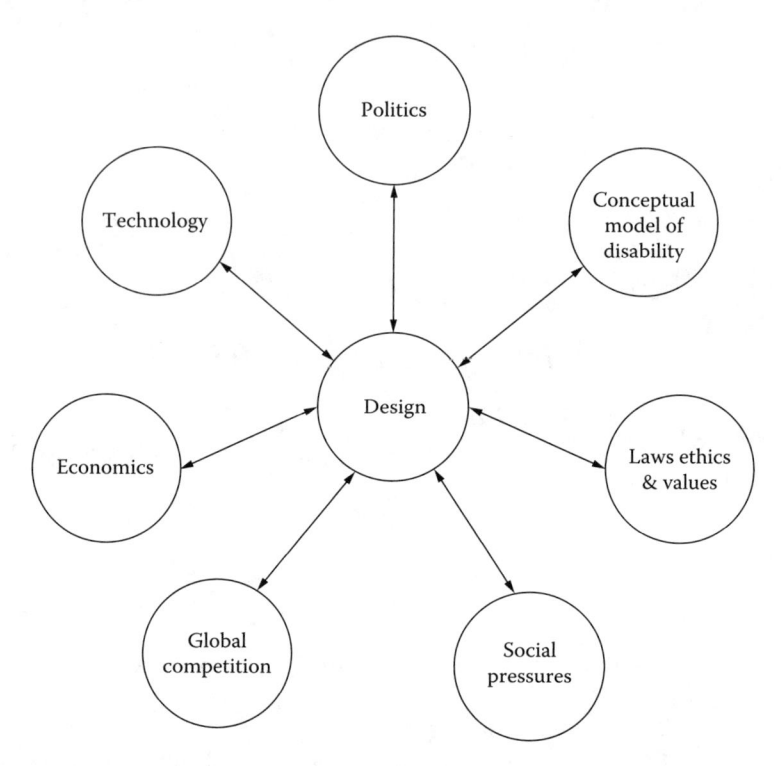

Figure 17.1 Design takes place within a complex set of linkages and interrelationships. This is Figure 1.1—but with a difference. Economics, politics, and social pressures have been added.

process is again placed at the center of this diagram to emphasize our focus on design and the fact that design takes place in a complex societal context.

Accessible and universal design principles and strategies derive from a simple fundamental premise: design entities to work with, support, and enhance human functioning. Design to human strengths and not human weaknesses. As the definition and applications evolve, it is important to remember this fundamental premise.

From here to where?

The ethical dimension of universal design goes back to its original conceptualization. "Architect Ron Mace coined the term in the early 1980s, as architectural requirements began to illustrate how access for people with disabilities usually meant better access for everyone" [1]. From this relatively humble beginning, universal design has morphed into a much more complicated idea.

Recall that in his article entitled "Creating Justice, Sustaining Life: The Role of Universal Design in the 21st Century," Weisman argues for an expanded "values-based framework for design decision making, based on an ethic of inclusiveness and interdependence that values and celebrates human difference and acknowledges humanity's debt to the earth" [2]. Weisman argues that it is time "to shift our focus from redressing human and environmental problems through remedial design to preventing problems through holistic design" [2]. Holistic design is another way of saying we need a broader systems view [3, 4]. From Mace to Weisman we have moved from universal design affording more accessible homes for everyone to universal design as a way to solve major world problems—this is quite a jump.

Thackara, in his book entitled *In the Bubble: Designing in a Complex World*, argues forcefully that, "If we can design our way into difficulty, we can design our way out" [5]. This resonates with Weisman's call for preemptive holistic design. However, Thackara tempers a strong preemptive holistic approach with a concern for unexpected consequences [5]. Our world is such a complex place that what we do in one context, while making sense at the time, has potentially devastating consequences as processes and systems evolve and grow, and as the scale of things change. Examples include the greenhouse effects of global warming and environmental pollution [5]. Addressing such issues as these requires designers to draw upon the principles of general systems theory that argue for a broader systems perspective [3, 4].

It is from this system's theory framework that Thackara argues the following: "If we can design our way into difficulty, we can design our way out." However, for this to work, we need to rethink what we mean by design [5]. Michael Douglas, artist, designer and Program Director of RMIT University, Melbourne, Australia, has been quoted as saying that "we need to pursue design practices which weave themselves through the social fabric without damaging it" [6, 7]. This is yet another way of saying that we need to better understand our design actions within a broader system's context, that is, within the linkages depicted in Figure 17.1.

Weisman, Thackara, and Douglas are just three voices, in an emerging chorus, stating the essential arguments for a new values-based universal design principle based on ecological concerns, but still consistent with the fundamental premise of designing entities to work with, support, and enhance human functioning. A clear articulation of such a principle has not yet evolved. Thackara states that in terms of addressing broad-based societal concerns, there are five types of "capital" with which society (designers) has to work; "natural, human, social, manufactured, and financial" [5]. How we manage this capital reflects the values and collective will of society suggested by the interaction of the elements in Figure 17.1.

This book has focused primarily on "manufactured" capital and secondarily on "human" capital. Manufactured capital in the broadest sense—products, services, and workplace processes, fixtures, and controls necessary to make and deliver products or services. Human capital in a more narrow sense of how people interact with and use the manufactured capital in terms of consumer products and services and work environments. Chapter 16 broadened the human perspective by exploring the ethical aspects of accessible and universal design. One conclusion from that discussion, in terms of this chapter, is that the manufactured capital can influence the perceived worth of human capital. Social capital can also influence the worth of human capital. The term "social capital" includes the broader social context within which we live and therefore covers not only social elements, but also political considerations.

Universal design from a systems perspective

Thackara urges that we need to design systems that are more "sustainable and light" (in terms of environmental and societal impacts) [5]. This formulation is moving closer to a values-based universal design principle in that sustainable and light are social values that influence how decision makers and designers approach and plan their projects. Formulating some form of "sustainable and light" as a transcending and integrating universal design principle would in turn create additional design strategies and provide an additional context in which to interpret the existing strategies.

Thackara presents three broad design strategies whose application could yield more sustainable and lighter systems [5]:

1. Minimize the waste of matter and energy.
2. Reduce the movement and distribution of goods.
3. Use more people and less matter.

These concepts can be viewed as a design principle hierarchy. The design of "sustainable and light" systems can be viewed as a values-based transcending and integrating principle supported by the three strategies. Each strategy has associated design substrategies, and such substrategies are in fact presented in Thackara's book, but not from this hierarchical perspective.

These new strategies can be related to the existing universal design principles as follows. The first two strategies are in harmony with the universal design principles related to *muda* (waste) reduction, the elimination of non-value-added activity (NVAA), and the reduction of variability—but from a global perspective. The third strategy—use more people and less matter—can be seen as supporting the first two strategies in that such an approach could make less demands on the environment and thereby be more sustainable. The third strategy can also be, arguably, promoted from the economic perspective of employing more people in a world with a rapidly growing population.

The evolution of universal design has and will continue to stimulate debate. Due to its original conceptualization by Mace, the inclusion of "equitability" as a transcending and integrating universal design principle has not been debated. However, as pressures mount to broaden the scope of universal design, we can expect continued serious debate about the inclusion of other values-based design principles—for example, ones dealing with "sustainable and light" systems or ecological concerns.

Regardless of where the debate goes, we as designers must still deal with the complexity suggested by Figure 17.1. The next section provides some examples of linkage complexity.

Examples of linkage complexity and the changing role of designers

Figure 17.1 diagrams that as the complex systems in our lives work to shape our lives, these same complex systems are, in turn, shaped by all the people who use and interact with them. As noted in previous sections, design is no longer a solitary act, but rather a collaborative activity. So many of the examples presented thus far illustrate that designers are moving from being "individual authors of objects, or buildings, to being facilitators of change among large groups of people" [5]. Recall (from Chapter 14) the impact of the design decisions by managers at Wal-Mart, McDonald's, Fuji Heavy Industries, Toyota, OXO, all the way down to small businesses such as Art for a Cause and Goodwill Industries. Also recall (from Chapter 15) the impact of Web designers on the creation of the World Wide Web following the W3C Web Accessibility Initiative design guidelines [8, 9]. Web designers are, in fact, the "facilitators of change among large groups of people."

We can examine the complexity and linkages suggested in Figure 17.1 by looking, once again, at the operations of Wal-Mart, McDonald's, Art for a Cause, and Goodwill Industries. With respect to the complexity of system interactions, Thackara could have been talking about Wal-Mart when he wrote the following: "Modern logistics, although undeniably impressive, is a smart answer to the wrong question. Logistics analyzes and optimizes the supply chain for a company and so creates value—but at a cost to the rest of us. In addition to the energy consumed, logistics consumes huge amounts of space and equipment. The total supply of short-term warehouse space increased from 6.1 to nearly 6.5 billion square feet between 1999 and 2000 alone. Faster movement of people or goods requires much greater public investment in transport systems. More logistics, however efficient, will make things worse if it is dedicated to shifting stuff further, faster. Rather than long-distance patterns of movement at accelerating speeds, local patterns or activity are a better destination" [6].

Wal-Mart's utilization of Six Sigma, *kaizen*, and other universal design strategies were, and are being, made from a conceptual model that is different than that presented by Thackara. If Wal-Mart had been making its system design decisions within a conceptual framework that had "sustainable and light" as one of the highest-level principles, its systems-level design approaches could have been very different. It is not the point of this discussion to speculate as to whether the results would have been "better," as such speculation is a value judgment, the discussion of which is beyond the scope of this book.

Complex interactions are also present with McDonald's mix of global suppliers and local suppliers, along with a corresponding mix of logistic and ecological consequences [10–14]. While McDonald's is a worldwide franchise operation, and the McDonald's logo is globally recognized, each country and locality also has unique food and merchandising elements. To serve such a diverse market with fresh produce, McDonald's has had to develop strong local supply networks [12–14]. Thus, while India's "Cold Chain" operation can move chickens and vegetables from "farm bulk to cooler in less than 90 minutes" [11–14], problems arise from exporting South American beef, chickens, and eggs; problems related to rain forest deforestation [10] to health concerns [13, 15], and to the humane treatment of the chickens [15]. McDonald's "Cold Chain" concept in India exemplifies Thackara's three strategies and is also representative of Six Sigma and *kaizen* strategies. The problems reported regarding the export of South American beef, chickens, and eggs are in addition to the logistic and storage concerns associated with such a large-scale operation. While the South American export strategy aimed at reducing costs and increasing the supply of necessary foodstuffs, these design objectives were offset by the ecological and social issues raised [12, 13, 15].

The scale of operations and management strategies of both Wal-Mart and McDonald's demands a global supply chain, yet local markets, pressures, and rising concerns regarding labor issues such as increased automation, outsourcing, downsizing, overseas investment, fuel consumption, global warming, and other ecological damage create pressure for new approaches [5, 13, 16]. Smaller-scale operations can explore and develop new approaches.

Art for a Cause exemplifies Thackara's three principles for more sustainable and light systems. By designing processes that support the employment of people with disabilities, the ETL/Art for a Cause collaboration focuses on using people as opposed to a more automated approach requiring more energy and fewer people. By establishing a global network of replicated operations, Art for a Cause will work to minimize the waste of matter and energy using human labor as opposed to automation, and will also work to reduce the movement and distribution of the goods produced. Each operation will receive the raw materials it requires and then distribute work to regional centers and finished product to regional markets. Each regional or national operation will exchange finished products that are unique to the supplier, but this international exchange can be reduced as workers at each location become skilled in replicating the new patterns and designs.

Goodwill Industries is also caught up in the global economy. Substantial portions of its donated clothes end up being baled and sold for recycling. Discussions with the retail manager of the West Bend, Wisconsin, store indicated that the baled material is sold as bulk to a buyer and distributor in Georgia. The distributor unbundles and sorts according to country— summer clothes to Africa or South America, winter clothes to eastern Europe, and wools are recycled in Italy for further use. Goodwill gets $0.04 to $0.06

per pound. The current baling and recycling process is ripe for redesign and overhaul.

This process raises questions such as the following. Can the clothes be recycled locally, rather than transported to Georgia and then around the world? If the baled clothes must go through Georgia, or some other intermediary, perhaps the supplying Goodwill sites should sort the clothing by seasonal use before baling? If Goodwill sites sorted the clothes, would the labor costs of sorting be shifted from the Georgia intermediary to Goodwill in terms of a higher payment per pound? These questions arise from the economic, political, and social context within which Goodwill operates, and the design decisions it makes at each store are influenced by such factors.

Goodwill operations have also entered the global e-commerce world: *"shopgoodwill.com"* is the first Internet auction site operated by a nonprofit organization [17]. In 1999, Goodwill Industries of Orange County, Santa Ana, California, was the first organization to start, and today most regional Goodwill organizations run a site in addition to the *"shopgoodwill.com"* operated by Goodwill International. Goodwill Industries International has 205 members with operations in 24 countries on 6 continents. Goodwill's online shopping categories include antiques, art, books and records, cameras, electronic equipment, clothing, collectibles, home décor, jewelry, musical instruments, and toys. The variety of items offered on *"shopgoodwill.com"* matches the breadth of larger commercial auction sites, with an average of 11,000 items listed on any given day [18]. "A clear leader in the online charity auction business, the site operates continuously, and gets more than 500,000 hits a day from all corners of the globe" [18].

The decision to enter the e-commerce market again reflects the technological, economic, political, and social contexts within which Goodwill operates. The system design decisions as to Web design, the logistics of inventory handling and shipping, and the auctioning and payment processes are all influenced by these broader contextual issues and linkages.

Art for a Cause and the Goodwill Industries operations described in Chapter 14 are, in many respects, the polar opposites of the McDonald's and Wal-Mart operations—not just in terms of size, scale of operations, and revenue, but also in terms of their use of accessible and universal design principles with respect to their respective employees. From a societal and political point of view, these distinctions sharpen the issues surrounding questions of how we should be using universal and accessible design.

Universal and accessible design in relation to society and politics

Universal design has gained notoriety as a result of laws mandating accessibility for people with disabilities, for example, to public buildings and public transportation systems [19], telecommunication products and services [20], and the broad spectrum of electronic and information technology [21].

Accessible design derives its definition from the laws that prescribe accessibility. The laws and associated guidelines [22] specify in a legal sense what is normative with respect to accessibility. In this way, accessible design principles, as formulated by laws and guidelines, shape society's assumptions about what is normative (that is, normal, acceptable, and expected).

While laws have gone a long way to address inequalities and promote equity for all citizens, it is time, as Weisman argues, "to move beyond the letter of the law to the spirit of the law; to shift our focus from redressing human and environmental problems through remedial design to preventing problems through holistic design" [2]. Weisman argues that universal design, more broadly defined than most current definitions, is a vehicle for moving in a new direction.

In his book entitled *Justice as Fairness* [22], Rawls, one of the world's most influential political philosophers, develops arguments proposing an ideal political structure, derived from a representative, participatory democracy embracing constitutional liberalism. The political structure proposed by Rawls can provide a new highway infrastructure for the universal design vehicle described by Weisman.

Every country and culture represents a unique environment and hence elicits a unique application of universal and accessible design practices. The primary focus of this book is design activities in the United States and other developed countries. Hence, for the purposes of this book, we limit the discussion to the ethical linkages and relationships between the principles and ideals inherent in a representative, participatory democracy embracing constitutional liberalism and accessible and universal design principles. It is the ethical principles underlying such democratic societies that provide the ethical imperatives for the promotion and advocacy of accessible and universal design.

It is important that we all agree on the terms used in the previous paragraph. The *New International Dictionary of the English Language Unabridged* defines a representative democracy as "a form of government in which the supreme power is vested in the people and exercised by them indirectly through a system of representation and delegated authority in which the people choose their officials and representatives at periodically held free elections" [23]. Zakaria, in his book entitled *The Future of Freedom* [24], notes that in 1900, "not a single country had what we would today consider a democracy: a government created by elections in which every adult citizen could vote." Today, Zakaria claims that 119 countries, or 62 percent of all countries, can be considered democratic.

Individual freedom and the ability and opportunity to fully participate as free and equal citizens in political debate and discussion is another hallmark of Western democracy [22]. Democracies thus characterized are termed "participatory democracies" [22]. Again, quoting Zakaria:

> For people in the West, democracy means "liberal democracy"—a political system marked not only by free and fair elections, but

also by the rule of law, separation of powers, and the protection of basic liberties of speech, assembly, religion, and property [24].

The term "liberal" is used in the 19th-century sense, sometimes called "classical liberalism." Liberalism in this sense is concerned with individual liberties: economic, political, and religious [24]. The term "liberal" is not used in the current political form—as in liberal versus conservative—and should not be construed to mean a "left"-leaning political orientation.

Hence, the United States is one of many countries that are a representative, participatory democracy embracing constitutional liberalism. Constitutional liberalism focuses on the government's goals. These goals are rooted in Western history "that seeks to protect an individual's autonomy and dignity against coercion, whatever the source—state, church, or society. The term marries two closely connected ideas. It is liberal because it draws on the philosophical strain, beginning with the Greeks and Romans, that emphasizes individual liberty. It is constitutional because it places the rule of law at the center of politics" [24, p. 19].

Laws are essential in that they form the ground rules for cooperation in our society. Laws are a set of mutually agreed-upon procedures, restrictions, and mandates that culturally diverse citizens collectively agree to follow. In a very real way, our laws reflect our collectively agreed-upon ethics and values. This brings us back to Rawls' ideal political structure derived from a representative, participatory democracy embracing constitutional liberalism.

Features of such societies include "measures assuring to all citizens adequate all-purpose means to make effective use of their liberties and opportunities" [25]. Justice as fairness regards "citizens as engaged in social cooperation, and hence as fully capable of doing so, and this over a complete life" [25]. Rawls argues that justice as fairness is an ideal theory, a theory that hypothesizes a well-ordered society and two fundamental principles of justice:

> Each person has the same indefeasible claim to a fully adequate scheme of equal basic liberties, which scheme is compatible with the same scheme of liberties for all; and
>
> Social and economic inequalities are to satisfy two conditions: first, they are to be attached to offices and positions open to all under conditions of fair equality of opportunity; and second, they are to be to the greatest benefit of the least-advantaged members of society (the difference principle) [22, pp. 42–43].

The first principle is consistent with a representative, participatory democracy embracing constitutional liberalism, such as in the United States. The second principle is where Rawls' ideas move more deeply into the realm of the ideal, in particular the difference principle and who are the "least-advantaged." According to Rawls, "In a well-ordered society where all citizens' equal basic rights and liberties and fair opportunities are secure, the least advantaged are those belonging to the income class with the lowest expectations. To say that inequalities in income and wealth are to be arranged for the greatest benefits of the least advantaged simply means that we are to compare schemes of cooperation by seeing how well off the least advantaged are under each scheme, and then to select the scheme under which the least advantaged are better off than they are under any other scheme" [22, pp. 59–60].

Disability rights advocates are quick to point out that disabled individuals experience inequalities in income and wealth and as a class have low expectations for economic improvement. If they, as a class, are not the group with lowest expectations, they are one of the lowest. Under Rawls' scheme, laws such as the ADA, Section 255 of the Telecommunications Act, and Section 508 of the Rehabilitation Act are examples of laws that satisfy the difference principle, in that they provide the greatest benefit to the least-advantaged members of society. These laws are designed to ensure that people with disabilities can in fact, as citizens, be fully engaged in social cooperation over a complete life.

As a citizen of a country, the country's laws and attitudes toward disability and the disabled play a dominant role in defining what is normative, what is expected, and what is valued. Returning to McBryde Johnson's interview with Berry Hawes [26], McBryde Johnson stated that people with impairments should be able to exercise more control over their environments and time. As an example, she cited the Medicaid system. Medicaid dollars are assigned to specific nursing home beds. She wants those dollars assigned to disabled individuals who would then decide how to spend the money. Why should a nurse, she asks, have to write an order for her to hire someone to change bedpans? To bathe her? To help her into a van? As McBryde Johnson says, "The system needs to let people make their own decisions" [26]. She goes on to say that "[t]he notion of a nurse or doctor to oversee this is silly. Our society has dehumanized people so much that this is considered OK, that we should just be happy that someone will do it" [26]. From her perspective, society places little value on the disabled, and she uses the term "dehumanized" to express this opinion.

McBryde Johnson's approach reflects the systems conceptual model of disability. People with impairments are citizens of the United States and have the right to be able to fully participate, as free and equal citizens, in all of society's activities. That includes lobbying for and securing the right to make economic and medical decisions for themselves. Society has the obligation to provide the infrastructure and resources necessary for this to happen—that is, to create a nondisabling environment.

In Rawls' formulation, society's expectations define those people who are considered least advantaged. As discussed in Chapter 16, the environment plays a significant role in determining society's expectations for individuals with disabilities. The blind office worker has different perceived value, depending on the availability of enabling technology. Therefore, if one adheres to the difference principle, then one must advocate for environments that are inherently the least disabling for as many people as possible.

If a society's laws establish accessibility and societal participation as the norm, then what is the role of universal design? In such a society, justice as fairness could arguably extend to viewing universal design as the only ethical design approach, one that explicitly sets out to allow maximum participation in the designed activity or with the designed entity. The justice as fairness model lends credence to Weisman's definition of universal design and his admonishment "to move beyond the letter of the law to the spirit of the law; to shift our focus from redressing human and environmental problems through remedial design to preventing problems through holistic design" [2].

Conclusions

Figure 17.1 diagrammatically captures the interrelatedness and complexity of the "design process." The concepts of universal and accessible design are evolving, as is evidenced by the varied definitions and conceptualizations presented in this book and as articulated in the mission statement of the International Association for Universal Design (IAUD). As part of its mission, the IAUD seeks to "carry out joint development and joint research on model cases beyond the framework of industry and business, seeking to manufacture products and create a social environment placing importance on the people-centered way of thinking, respecting the humanity of each person. We will take into account not only the products and services that are related to UD, but also the elements that influence the way we live, including creating the environment, how we spend the leisure time, inheritance of culture, and education" [27].

The universal and accessible design principles presented in Sections II and III derive from more narrowly conceived definitions of universal and accessible design. Given the evolving nature of universal design, it is natural to assume that more principles will be added over time—for example, principles that more explicitly deal with sustainability and lightness.

How these principles are applied will depend on how we as a society express our collective values in our laws. Recent court and ballot initiatives successfully promoted legal action to curtail or suspend affirmative action in employment and college recruitment and admissions [28–30], and the courts have a history of defining the limits of the ADA [31–34]. The social values of fairness and justice are at the heart of all these legal issues; and as these social values are also evolving, they will eventually find expression in laws. These laws, in turn, will then influence and mandate design considerations and decisions.

Design is a fundamental human activity [35–40]. For as stated in Chapter 2, "design is a thought process comprising the creation of an entity" [35]. As such, all of us are designers.

I want to restate that accessible and universal design principles and strategies derive from a simple fundamental premise: design entities to work with, support, and enhance human functioning. Design to human strengths and not human weaknesses. As the definition and applications evolve it is important to remember this fundamental premise. With respect to how we as designers understand our roles in the design process depicted in Figure 17.1, I want to end with a quote from Rabindranath Tagore's book *Fireflies* [41]:

> I miss the meaning of my own part
> in the play of life
> because I know not of the parts
> that others play.

References

1. IDSA, Industrial Designers Society of America Universal Design, http://www.idsa.org/whatsnew/sections/udidsa/noframes.htm, 2004.
2. L. K. Weisman, Creating Justice, Sustaining Life: The Role of Universal Design in the 21st Century, presented at Adaptive Environments Center 20th Anniversary Celebration, The Computer Museum, Boston, April 10, 1999.
3. L. von Bertalanffy, *General System Theory*. New York: Brazillerr, 1969.
4. E. Laszlo, *Introduction to Systems Philosophy*. New York: Harper Torchbooks, 1972.
5. J. Thackara, *In the Bubble: Designing in a Complex World*. Cambridge, MA: MIT Press, 2005.
6. L. Trachtman, User-Centered Design: Principles to Live By, *Assistive Technol.*, 10, 1–2, 1998.
7. M. Douglas, Imagining Transports of Sustainability, presented at *Humanities Conference: International Conference on New Direction in Humanities*, University of Agean, Island of Rhodes, Greece, 2003.
8. WAI, Checklist of Checkpoints for Web Content Accessibility Guidelines 1.0, http://www.w3.org/TR/WCAG10/full-checklist.html, 1999.
9. WAI, Web Accessibility Initiative (WAI) — WAI Homepage, http://www.w3.org/WAI/, 2/5/02.
10. I. Forests.org, McDonald's Linked to Rainforest Destruction, http://forests.org/archive/general/macfore.htm, The McLibel Trial, World Rainforest Report Rainforest Information Centre, 1996.
11. ICFAI: Case Studies, McDonald's Food Chain: India, *ICFAI Center for Management Research Asia's Largest Online Collection of Management Case Studies*, Case Study Number: OPER001, 2002.
12. C. Klopper and A. Joyce, Forecasting and Replenishment Solutions: McDonald's Europe Confirms Successful Implementation of Manugistics' Advanced Forecasting and Replenishment Solutions, *Mulberry Marketing Communications*, http://www.e-consultancy.com/news-blog/359854/mcdonald-s-europe-confirms-successful-implementation-of-manugistics-advanced-forecasting-and-replenishment-solutions.html?keywords=supply+chain, 2004.
13. E. Schlosser, *Fast Food Nation*. New York: Houghton Mifflin Co., 2004.
14. McDonalds's India, McDonald's India's Cold Chain, http://www.mcdonaldsindia.com/aboutus/presskit/04%20Cold%20Chain.pdf, 2004.

15. D. Meadows, Consumer Power Reforms Chicken Factories — But Not Enough, http://www.pcdf.org/meadows/antibiotics.html, *The Global Citizen,* 2000.
16. J. Dicker, *The United States of Wal Mart.* New York: Jeremy P. Tarcher/Penguin, 2005.
17. Goodwill Industries International, shopgoodwill.com is the first Internet auction site operated by a nonprofit organization, http://www.goodwill.org/page/guest/about/howweoperate/Shopping/onlineauction, 2005.
18. Goodwill Industries of Southeastern Wisconsin, Inc., & Goodwill Industries of Metropolitan Chicago, Inc., this site is home to Goodwill Industries of Southeastern Wisconsin and Metropolitan Chicago, http://www.goodwillsew.com/, 2006.
19. Access Board, Americans with Disabilities Act Accessibility Requirements, http://www.access-board.gov/bfdg/adares.htm., 1999.
20. Access Board, *Telecommunications Act Accessibility Guidelines: Final Rule.* Washington, D.C.: *Federal Register,* 36 CFR Part 1193, 1998.
21. Access Board, *Electronic and Information Technology Accessibility Standards.* Washington, D.C.: *Federal Register,* 36 CFR Part 1194, 2000.
22. J. Rawls, *Justice as Fairness.* Cambridge, MA: The Belknap Press of Harvard University Press, 2001.
23. Merriam Webster, *Webster's Third New International Dictionary of the English Language Unabridged.* Chicago, IL: William Benton, 1966.
24. F. Zakaria, *The Future of Freedom: Liberal Democracy at Home and Abroad.* New York: W. W. Norton & Company, 2003.
25. J. Rawls, *Political Liberalism.* New York: Columbia University Press, 1996.
26. J. Berry Hawes, Harriet McBryde Johnson, in *The Post and Courier.* Charleston, 2003.
27. IAUD, International Association for Universal Design, http://www.iaud.net/en/member/index.html, 2004.
28. P. Egan, M. Schultz, and C. Jun, Affirmative Action Backlash: Group Files Suit, U-M, Others May Challenge Ban, *The Detroit News:* http://www.detnews.com/apps/pbcs.dll/article?AID=/20061109/POLITICS01/611090379, Thursday, November 9, 2006.
29. M. Russell, Backlash, The Political Economy, and Structural Exclusion, *Berkeley J. Emp. & Lab*, 21, 335–340, 2000.
30. S. Sturm and L. Guinier, The Future of Affirmative Action, *Boston Review,* http://bostonreview.net/BR25.6/sturm.html, originally published in the December 2000/January 2001 issue of *Boston Review.*
31.] Reason on Line, Disability Debate, *reasononline:* http://oldsite.reason.com/bi-ada.shtml, 1999.
32. D. Ackman, Supreme Court Defines Disability, *Forbes.com,* 2002. http:www.forbes.com/2002/01/09/0109 top news.
33. Bazelon Center for Mental Health Law, Supreme Court Decides ADA Case on Definition of Disability, Bazelon Center for Mental Health Law, 2002. http:www. bazelon.org/williams.html.
34. The Center for an Accessible Society, Supreme Court Limits Americans with Disabilities Act in Williams Case, *The Center for an Accessible Society,* http://www.accessiblesociety.org/topics/ada/williams2.htm, 2002.
35. W. R. Miller, The Definition of Design, http://www.tcdc.com/dphils/dphil1.htm, 1996.

36. D. J., Boorstin, *The Discoverers*. New York: Vintage Books, a division of Random House, 1983.
37. S. Yelavich and S. Doyle, *Design for Life*, Washington, D.C.: Cooper-Hewitt, National Design Museum, Smithsonian Institute, 1997.
38. C. Dym, L. and P. Little, *Engineering Design*. New York: John Wiley & Sons, 2000.
39. D. A. Norman, *The Design of Everyday Things*. Originally published in hardcover by Basic Books, 1988: Basic Books – Member of the Perseus Book Group, 2002.
40. R. F. Erlandson, Universal Design for Learning: Curriculum, Technology, and Accessibility, in *ED-MEDIA 2002 World Conference on Educational Multimedia, Hypermedia & Telecommunication*. June 26–29, 2002, Denver, CO: Invited Paper, 2002.
41. R. Tagore, *Fireflies*. New York: Macmillan Publishing, 1928.

Index

A

AARP, agreement between Home Depot and, 7
ABA, see Architectural Barriers Act
Absolute judgment, 92
Access Board, 11
　accessibility guidelines of, 18
　accessible design technical assistance, 54
　creation of, 6
　federal departments of, 38
　first priority of, 38
　guidelines and standards, support for, 54
　mandated standards published by, 37
　published guidelines and standards, 39–41
　resource materials, 215
　responsibilities of, 11, 38
　Section 508 requirements and, 207
　standardized format for design assistance documents, 55, 56–57
　website of, 214
Accessibility Forum, 11
　focus of, 55
　member organizations, 57
　reason for creation of, 54
　Section 508 Buy Wizard, 57
　size of, 57
Accessibility laws, publicity on, 10
Accessible design, 37–61, see also Universal design/accessible design/adaptable design
　Access Board, 37–41, 54
　accessibility forum, 54–57
　accessibility laws and, 238
　Center for Accessible Housing definition of, 18
　definition of, 6, 18
　examples of Web-based material on, 11
　laws and accessibility, 42–54
　　Americans with Disabilities Act (1990), 42–46
　　Architectural Barriers Act (1968) as amended 42 U.S.C. § 4151 et seq, 42
　　compatibility, 49–50
　　electronic and information technology, 51–54
　　Telecommunications Act (1996), 46–49
　strategies, fundamental premise of, 67
Accessible Design Resource Center (ADRC), 11, 215
Accessible Technology in Today's Business: Case Studies for Success, 225
Accessible Telecommunications Product Design Tutorial, 54, 56–57
Accidents, errors and, 143
Accommodation, distinction between *poka-yoke* intervention and, 132
ADA, see Americans with Disabilities Act
ADA Accessibility Guidelines (ADAAG), 20, 39, 44, 74, 79
　Accessible Transient Lodging of, 85
　amendments
　　building elements for children, 39
　　over-the-road buses, 40
　　play areas, 40
　　recreation facilities, 40
　　state and local government facilities, 39
　ergonomic considerations, 79
　transportation vehicles, 40
ADAAG, see ADA Accessibility Guidelines
Adaptable design, see also Universal design/accessible design/adaptable design
　accommodations of, 18
　definition of, 18

245

example of, 20, 21
  most common example of, 19
Administrative controls, 76, 127
Adobe Systems, Inc., 57
ADRC, *see* Accessible Design Resource
    Center
AESC, *see* American Engineering Standards
    Committee
Affordance(s)
  description of, 97
  effectiveness of, 104
  object design determined by, 99
Agile manufacturing, 125
Agilent Technologies, 121
Agile system
  design, 153–154
  technology, example of, 125
Airports, reduction of common cause
    variability at, 172
AlphaSmart®, 184, 185
Alzheimer's patient, GPS technology and,
    182
Ambient noise, 85
American Engineering Standards Committee
    (AESC), 161
American National Standards Institute
    (ANSI), 50, 161
Americans with Disabilities Act (ADA), 7, 18,
    37, 42
  curb cuts mandated by, 20
  difference principle and, 240
  Title I, 42
  Titles II and III, 44
  Title IV, 46
  Web accessibility and, 46
America Online, Inc., 57
ANSI, *see* American National Standards
    Institute
Anthropometry, personal perspective on, 77
Apple
  accessibility features in operating systems
    of, 15
  needs of disabled community addressed
    by, 16
Appliance
  general-purpose device versus, 150
  plug-and-play function of, 151
Architectural Barriers Act (ABA), 7, 37, 42
Architectural and Transportation Barriers
    Compliance Board, *see* Access
    Board
Art for a Cause, 192, 236, 237
Assembly jobs, NVAA classification of, 147

Assembly operations, *poka-yoke* devices at,
    137
Assistive technology (AT), 120
  device, definition of, 43
  E&IT and, 52
  flexibility and, 129
  service, definition of, 43
Assistive Technology Industry Association
    (ATIA), 123
AT, *see* Assistive technology
ATIA, *see* Assistive Technology Industry
    Association
ATMs, *see* Automated teller machines
"At the source" error-proofing, 134
AT&T™ 9350 Digital Answering System
    Speakerphone, 92
Auditory information, availability of, 85
Auditory prompts, 84
Authoring tool(s)
  description of, 213
  explosion of, 227
Automated teller machines (ATMs), 79, 92
Automatic door openers, 20, 180
Automobile(s)
  adjustability, 20, 119, 180
  assembly of components on, 93
  dynamics, testing of, 16
  environmental adjustability in, 123
  inherent variability of, 159
  manufacturers, use of anthropometric
    data by, 78
Automotive designers, knowledge built into
    internal environment, 96

B

Baby boomer(s)
  aging of, 80
  market, 7
Background noise, 85
"Before improvement" processes, 154
Bell South, 57
Binoculars, visual stabilization circuitry in,
    183
Birthways Empathy Belly, 78
Bluetooth voice recognition kits, 122
Body size and shape, ergonomics and, 77
Booz Allen & Hamilton, 57
Bose®Wave Radio/CD, 89, 109, 112
Braille writing, elevator controls with, 18
Brake-fitting fixture, process redesign of, 135,
    137
Breakdowns, inefficiencies of, 149
Bush, George W., 3

Index

C

CAD, *see* Computer-aided design
California car, 51
Cameras, visual stabilization circuitry in, 183
Carrying strategy, torque and, 73–74
CAST, *see* Center for Applied Special Technology
Catastrophic errors, 17
Center for Accessible Housing, 6
    definition of accessible design, 18
    definition of universal design by, 17
Center for Applied Special Technology (CAST), 11, 114, 127
Center for Universal Design, The, 11
Choice, paradox of, 148, 153
Cingular Wireless LLC, 57
Classical liberalism, 239
Closed captioning, 46, 184
Code requirements, 18
Cognitive demands, 95
Cognitive design strategies, training and educational issues, 114
Cognitive impairments, workers having, 44
Cognitively sound design, 95–117
    affordance, 97–99
    build knowledge into designed entity or environment, 97
    complexity, 106–115
    constraints, 103
    design strategies, 97
    discussion, 95–97
    education and training, 114
    feedback, 103–104
    language, 104
    mappings and representation, 99–103
    principle, 95
    provide messages in language and format that people will understand, 105–106
        appropriate level of language, 105–106
        appropriate style of language, 105
        language that users understand, 105
        universally understood icons, symbols, or pictures for communications, 106
    reduce operational complexity of entity, 106–113
    use universal design for learning strategies developed by CAST for training or educational activities, 114–115
Coin counting fixtures, 136

Common attributes, ergonomic design and, 73
Common cause, 157
Common cause variability
    airports, 172
    broom balancing example, 158
    design of entities to reduce, 157
    establishment of standards to reduce, 159
    levels, 159
    meter reading, 171
    parking process, 170
    reduction of, 166
    strategies for reducing, 160
    use of visual controls in reducing, 169
Communication, use of human senses as, 88
Compatibility
    ensuring, 123
    law mandating, 49
Competition, global, 4
Computer
    access, student, 141
    -aided design (CAD), 166, 187
    mouse, double-click speed of, 125
    operating systems, testing of, 16
    programs, word prediction associated with, 140
Concept mapping, 115
Conceptual model(s)
    affordances and, 97
    development of, 30
    global, 216
    mappings and, 99
    reality and, 99
    system elements, 29–30
    WHO-ICF, 31, 32
Constitutional liberalism, 239
Constraints
    cultural, 103
    equitability and, 179
    reason for using, 103
Consumer product(s)
    design, *muda* elimination in, 148
    volume controls on, 89
Consummate Design Center, 15
Corporate conduct, Securities and Exchange Commission investigation of, 221
Creform® Corporation, 125, 126
Crossing the Chasm, 151
Cultural constraints, 103
Curb cuts, 20, 180
Customer feedback, automobile design and, 16
CuteTools™, 192, 193

D

Danfoss Company, 106
Dassault Systemes
    CAD software by, 187
    enabling technology by, 166
    three-dimensional visualization tools, 197
DCMI, *see* Dublin Core Metadata Initiative
Decision trees, 108
Dehumanized society, 240
Dell Computer Corporation, 57
Delmia, ENVISION/ERGO™, 79
Democracy, hallmark of Western, 238
Department of Defense (DOD), 183
Design
    decision making, values-based framework for, 228, 232
    definition of, 15, 17
    ergonomic, 68
    multimedia approach to, 53
    process, 3–13
        building equity into, 185
        complexity of, 232, 241
        definitions, 5–6
        laws, 6–7
        linkages, 3–5, 231
        mainstreaming issues, 10–11
        market and ethical imperatives for accessible and universal design, 7–10
    relationships among types of, 19
    requirements, standards and, 187
    software, 166
    solution, cookbook type of, 83
    subcategories, 17
"Design for all" principles, 11
Design for assembly (DFA), 93
Design Basics, 124
Design of Everyday Things, The, 99, 133
Device(s)
    common cause variability inherent in, 162
    feedback, 104
    general-purpose, 150
    modes of, 111
    plug-and-play function, 151
    poka-yoke, 137, 148
DFA, *see* Design for assembly
Diebold, 57
Difference principle, 240
Diffuse project, 211
Digital cell phone compatibility, 50
Disability(ies)
    awareness training, 222
    community, civil rights movements of, 3
    country's attitudes toward, 240
    creation of job opportunities for people with, 43
    definition of, 30
    ethics of, 221
    global conceptual model of, 216
    market, design strategy for, 9
    native endowments and, 224
    relationship between environmental factors and, 31
    society expectations and, 240
    systems model of, 34
    WHO conceptual model of, 225
Disability and design, 29–36
    accessible design, universal design, and conceptual models of disability, 34–35
    conceptual models, 29–30, 35
    definition of disability, 30–32
    design implications associated with WHO-ICF conceptual model, 32–34
        medical model, 33
        social model, 33
        systems model, 34
    enabling and disabling environmental factors, 32
DMADV process design, 164
DOD, *see* Department of Defense
DOD TechMatch, 183
DO-IT project, 215
Don Johnston, 57
Doors, design standards for, 45
Dublin Core Metadata Initiative (DCMI), 212
Dutch Association of Bioethics, 223

E

Ecological concerns, universal design principle based on, 233
E-commerce global economy, 8, 193, 197
Efficiency
    NVAA and, 70
    principle, 150
Efficient design, 147–155
    appliances versus general-purpose devices, 150–151
    design strategies, 149–150
        quality control and reliability engineering techniques, 149–150
        simple and easy to use as possible, 150
    discussion, 147–149
    principle, 147
    process and service considerations, 152–154

avoid complexity in that it leads to NVAA, 152–153
use task analysis techniques to identify tasks or activities that can be eliminated or redesigned so as to reduce or eliminate NVAA, 153–154
E&IT, *see* Electronic and information technology
Electric shock, 87
Electronic and information technology (E&IT), 51, 79
  assistive technology and, 52
  definition of, 51
  final standards for, 41
  products, growing importance of, 216
E-mail, 102
Emergency signals, 83, 90
Enabling Technologies Laboratory (ETL), 10, 122, 138, 193
Enabling technology, 166
Engagement, levels of, 114
Engineering controls, 76
Entity(ies)
  aesthetically pleasing, 180
  building of knowledge into, 97
  creation of, 16, 17
  designing flexibility into, 119
  marketing of, 184
  operational complexity of, 106
  user reaction time and, 108
  workplace, 188
Environment
  examples of knowledge built into, 96
  influence of personal factors on functioning in, 30
  interaction with, systems model and, 34
  level of participation and, 31
  social model importance of, 35
  systems model characterization of, 34
Environmental barriers, removal of, 70
Environmental enablers, examples of, 32
Environmental factors
  enabling and disabling, 32
  process failure and, 163
ENVISION/ERGO™, 79
Equitability, 70
  constraints imposed by, 69
  value of, 69
Equitable design, 179–186
  design strategies, 180–185
    aesthetically pleasing, 180–182
    age and context appropriate, 180
    competitively priced, 182–184

market entity for broad demographic and socioeconomic base, 184–185
discussion, 179–180
principle, 179
transcending and integrating, 185
Ergonomically bad design, 18
Ergonomically sound design, 73–82
  design strategies, 76–81
    avoid ergonomic risk factors, 76–77
    ease of use, 80–81
    wide range of body sizes and shapes, 77–80
  discussion, 73–76
  principle, 73
Ergonomic design, definition of, 73
Ergonomic engineering, design strategies for, 81
Ergonomics
  definition of, 68
  OSHA standards definition of, 76
Ergonomics Society, 68
Error(s)
  catastrophic, 17
  categories of, 133
  consequences of making, 167
  correction of, 143
  mistakes causing, 139
  -prone processes, judgments as, 170
  rates, production rates vs., 44
  recovery, 138, 149
  -reducing tools, office job, 132
  use of sensor to detect, 134
Error-managed design, 131–145
  design strategies, 133–140
    mistakes, 139–140
    slips, 133
    three-staged approach, 133–139
  discussion, 131–132
  error-proofing for training and educational processes, 140–143
  principle, 131
Error-proofing
  "at the source", 134
  examples using warning, 135
  most effective approach to, 133
  process redesign with goal of, 137
  safety risks eliminated by, 198
  strategies, process-oriented, 133
  talking scale system and, 9
  task, 70
Ethical code, Southwestern Bell Communications, 228
Ethics and universal design, 221–230
  ethics of disability, 221–224

native endowments, environment, and
        disability, 224–226
    universal design, 226–229
ETL, see Enabling Technologies Laboratory
EUROPA, 121, 122
European Commission's Directorate, General
        Education and Culture, 22
European Concept for Accessibility Network,
        24, 227
European Union
    integration of universal design material in
        higher education in, 10
    website, 121
Eyeglasses, universal accessibility of, 182

F

Fast food
    companies, demands on suppliers by, 192
    industry, lean production in, 190
    preparation, Speedee Service System of,
        191
FCC, see Federal Communications
        Commission
FDA, see U.S. Food and Drug Administration
Fechner's law, 87
    examples of, 87
    physical stimuli and, 87
    signal detection expressed by, 86, 90
Federal Communications Commission
        (FCC), 50
Feedback
    device, 104
    signaling and, 119
    use of, 103
Fireflies, 242
Fitts' law, 92, 93
Five Ss, lean production using, 167
Flexibility
    assistive technology and, 129
    design challenge with, 128
    design overhead with, 128
Flexible design, 119–130
    build flexibility into service delivery
        systems and work processes,
        126–127
    design challenge, 128
    discussion, 119–120
    principle, 119
    provide adjustability and mobility,
        123–126
        adjustable response times, 124–125
        agile system design strategies, 125–126

ergonomic and environmental
        adjustability, 123–124
    perceptual adjustability, 124
provide user with choices, 120–123
    assistive technologies used with
        designed entity, 122–123
    choice of language, 121–122
    choice of mode for communication,
        122
Food
    designers, concern of, 88
    quality, multisensory phenomenon of, 88
Forcing functions, 103, 133, 134
Ford Motor Company, 57
    pregnancy simulator, 78
    Third Age Suit, 7, 16, 78
Four-step process, reliability of, 165
Freeplay radio, 22–23, 151
Fuji Heavy Industries Ltd., universal design
        at, 188
Functional performance criteria, 52, 53
*Future of the Disabled in Liberal Society: An
        Ethical Analysis, The*, 223
Future of Freedom, The, 238
Fuzzy sets, theory of, 17

G

Gas pump nozzles, shut-off valves on, 134
Gateway Computer, 57
General Motors Corp., car designed for
        pregnant women, 78
General-purpose device, appliance versus,
        150
General Services Administration (GSA), 11,
        55, 214
Genetic testing, 223
Global competition, law and, 4
Global economy, e-commerce, 193, 197
Global market(s)
    competition among, 23
    universal design in, 22
Global marketing, universal design and, 8
Global vocabulary, societal influences
        contributing to, 102
Goodwill Industries, 237
    Great Lakes Naval Station–Food Service
            Operations, 196
    retail outlet facility, 195
    use of *kaizen* method by, 193
Google site, 108–109
GPS technology, 182
Graph theory, 107–108
GSA, see General Services Administration

Index

H

Hard-of-hearing persons, telephone use by, 20–21
HDTVs, 128
Hearing aid compatibility, 50
Henry Ford's Greenfield Village, The, 75
Hewlett Packard, 57, 140
Hick's law, example of, 108
Hierarchical structure of universal design principles, 67–71
   design philosophy, 67
   hierarchical constraints, 70–71
   principles targeted at person, 68–69
   principles targeted at process, 69
   transcending, integrating principle, 69–70
   universal design principles, 67–68
Higher education, integration of universal design material in, 10
HiSoftware AccVerify SE™, 213
Home(s)
   automation system, flexibility in, 120
   average size of, 80
   health care, remote sensor technology and, 120
   universally designed, 80
Home Depot, agreement between AARP and, 7
Housing, evolving, 124
Human capital, 233
Human ear, infrared image of, 100
Human factors principles, 68, 70
Human judgment, information processing capabilities and, 90
Human physiology, 74

I

IAUD, see International Association for Universal Design
IBM, 30, 57
ICIDH, see International Classification of Impairments, Disabilities, and Handicaps
Icons, universally understood, 106
IEA, see International Ergonomics Association
IEEE, see Institute for Electrical and Electronic Engineers
IETF, see Internet Engineering Task Force
Image of possibility, 16
In the Bubble: Designing in a Complex World, 233
Information

accessibility, ADA and, 46
activities, Section 508–specified, 52
auditory, availability of, 85
exchange infrastructure, Internet as part of, 202
Internet, 210–211
multiple means of representing, 115
processes, neural networks specialized for, 114
processing capabilities, human judgment and, 90
processing channel capacity, 101
retrieval, 53
theory, 17
visual, availability of, 84
workplace, 85
Information Society Technologies (IST), 211
Information Technology Technical Assistance and Training Center, 202
Inspiration initiative, 189
Institute for Electrical and Electronic Engineers (IEEE), 11, 161
   Computer Society, 57
   Reliability Society, 163
Interface display, shrinking, 150
International Association for Universal Design (IAUD), 197, 241
International Classification of Impairments, Disabilities, and Handicaps (ICIDH), 33
International Ergonomics Association (IEA), 68
International Standards Organization (ISO), 132, 161
Internet
   ADA accessibility requirements to use, 46
   -based activities, organizations promoting standards for, 212
   connection, feedback and, 103
   design community, procedures of, 201
   information, Web-based, 210–211
Internet Engineering Task Force (IETF), 212
Intervention techniques, categories of, 76
Intranet information, Web-based, 210–211
Inventory management system, Wal-Mart's, 191
Invisible Computer, The, 150
ISO, see International Standards Organization
IST, see Information Society Technologies

J

Jaws©, 30
JND, see Just noticeable difference

Job
　designs, transportable, 198
　essential functions, 43, 154
　opportunities, creation of for people with disabilities, 43
　redefinitions, 135, 154
Justice as Fairness, 238
Just-in-time inventory, 191
Just noticeable difference (JND), 86, 87

K

Kaizen
　muda elimination in, 147, 148
　strategy
　　fundamental, 132
　　Goodwill Industries use of, 193
　　McDonald's, 192
　　Wal-Mart, 235
Katas, 169
KISS, 115
Kitchen range burner controls, mapping and, 101
Knowledge, building of into environment, 96, 97

L

Language
　message comprehension and, 104
　strategy, 105
Laptop computer system, warning message on, 135
Law(s)
　basis of, 3
　definition of, 239
　disciplines impacted by, 6–7
　global competition and, 4
Lean production, 166
　error-proofing and, 132
　pressures imposed by certification requirements, 132
　strategies, 167
　Subaru manufacturing plant using, 188
Leisure activities, challenging, 107
Liberal democracy, 238
Liberalism, 239
Lifting strategy, torque and, 73–74
Light curtain sensing system, 137, 138
LilTrak 255 pedometer, 151
Limited manual dexterity, device operable with, 75, 81
Living environment, customization of, 124
Lockheed Martin, 57

M

Macromedia Dreamweaver, 214
Mandela, Nelson, 22
Manufactured capital, 233
Manufacturing
　designing for cost-effective, 187
　industry, creativity in, 189
　jobs, NVAA classification of, 147
Mapping(s)
　conceptual models and, 99
　definition of, 99
　example, 101
　influences contributing to, 102
　knowledge-building process and, 100
Marginally literate Americans, 105
Market share, global economy and, 8
Marshall Space Flight Center (MSFC), 184
McDonald's, 237
　"Cold Chain" concept, 236
　kaizen strategies used by, 192
　local supply networks, 236
　workplace organization of, 192
Medicaid system, 240
Medical model, view of disability in, 33
Memory-related human performance capabilities, 110
Meter reading, error-prone processes and, 170
Microprocessor technology, cameras containing, 183
Microsoft
　Accessibility, 11
　accessibility features in operating systems of, 15
　FrontPage, 213
　HiSoftware and, 213
　needs of disabled community addressed by, 16
　Visual Studio, 140
MindJet®, 115
Mistakes
　errors resulting from, 139, 140
　learning by making, 140, 141
Mobile phone, flip-open, 111
Model(s)
　conceptual, 31, 32
　　affordances and, 97
　　development of, 30
　　global, 216
　　mappings and, 99
　　reality and, 99
　　system elements, 29–30
　consumer buying behavior, 148
　Hick's law, 108

medical, view of disability in, 33
social, 33
    importance of environment in, 35
    view of disability in, 33, 43
systems, view of disability in, 34
WHO-ICF, 31, 32, 54, 216, 226
Modes, device, 111
Morse code, 91
Motorola, 57
    Hands-Free Neckloop accessory, 123
    Six Sigma, 163
MSDs, see Musculoskeletal disorders
MSFC, see Marshall Space Flight Center
Muda, meaning of, 147
Muda elimination, see Efficient design
Multisensory communication options, 84
Musculoskeletal disorders (MSDs), 76
Museum of Modern Art, The, Design Collection of, 5, 182

N

Nano-devices, 150
NASA technology, 183, 184
National Association of Home Builders, 80
National Center for Accessible Media (NCAM), 213
National Center for Health Statistics (NCHS), 9
National Institute for Occupational Safety and Health (NIOSH), 79
National Institute of Standards and Technology (NIST), 159
Native endowments, disability and, 224
Natural mappings, 99
NCAM, see National Center for Accessible Media
NCHS, see National Center for Health Statistics
New Freedom Initiative, 3, 35
Niche markets, 10
NIOSH, see National Institute for Occupational Safety and Health
NIST, see National Institute of Standards and Technology
Non-value-added activity (NVAA), 70, 147, 162, 234
    classification, 147
    complexity reduced by eliminating, 162
    computer-based online help systems, 153
    elimination of, 234
    ergonomic principles reducing, 147
    inspection of process operations as, 154
    phone service systems exemplifying, 152
    reduction of, 149, 166, 190
    service applications exhibiting, 152
    task analysis techniques used to reduce, 153
Northrop Grumman IT, 57
NVAA, see Non-value-added activity

O

Object design, affordance and, 99
Occupational Safety and Health Administration (OSHA), 76
    definition of ergonomics, 76
    ergonomics standards document, administrative controls in, 127
    requirements, worker safety and, 198
    workplace requirements, 79
Occupational therapy, 32
Office jobs, error-reducing tools in, 132
Online help systems, computer-based, 153
OSHA, see Occupational Safety and Health Administration
OTRBs, see Over-the-road buses
Over-the-road buses (OTRBs), 40
OXO International, Ltd.
    design awards, 5
    design philosophy of, 180
    devices produced by, 181
    marketing success of, 5
    principles of universal design for, 8, 197
    product lines, 4

P

Pain signals, 87
Palm-held devices, prediction strategy used by, 140
Panasonic, 57
Paradox of Choice, 148
Parking process, common cause variability in, 170
Participatory democracies, 238
Perceptible design, 83–94
    design entities so user can accurately acquire target, 92–94
    design signals to maximize signal-to-noise ratio, 85–91
    design strategies, 84–94
    discussion, 83–84
    keep signaling structure and content as simple as possible, 91–92
    principle, 83

provide multisensory options for communications between person and process or product, 84–85
Perceptual issue, most obvious, 84
Phone-based customer support system, flexibility of, 121
Phone service systems, NVAAs of, 152
Physical constraints, error-proofing using, 133
Physical therapy, 32
Physiological impairments, perceptual abilities and, 94
PIN numbers, 79, 110
Poka-yoke
  devices, 137, 138
  intervention, 132
Poka-yoke: Improving Product Quality by Preventing Defects, 133
Political actions, 34
Political perspective, *see* Social and political perspective, universal and accessible design from
Politics, universal design and, 237
Possibility, link between problem and, 15–16
Predictable design, *see* Stable and predictable design
Prediction strategy, 140
Pregnancy suit, 78
Pregnant women, automobiles designed for, 78
Probability theory
  process redesign and, 164
  process reliability and, 165
Process
  building of knowledge into, 137
  definition of, 69
  engineering, unstable system in, 163
  error-proofing of, 69
  four-step, 165
  -oriented error-proofing strategies, 133
  -oriented universal design principle, 143
  principles, 68
  reason for designing, 131
  redesign, DMAIC approach to, 164
Product(s)
  common cause variability inherent in, 162
  inspection, 90, 154
  value, work steps and, 149
Prosthetics, compatibility of controls with, 49
Public building(s)
  automatic door openers in, 180
  interior lighting of, 85
  muda elimination in design of, 148
  ramps to existing, 21
  technical requirements in design of, 45
Public kiosks, 89
Public streetlights, 85

Q

QS, *see* Quality System
Qualified individual with disability, ADA definition of, 42–43
Quality control
  methods, Goodwill Industries use of, 196
  programs, error-proofing and, 132
  techniques
    design of, 149
    reducing common cause variability using, 162, 163
Quality System (QS), 132

R

Radio frequency identification (RFID) tags, 191
Rainbow medical color palette, 100
Ramps, 21, 74
Range-of-motion, 77
Reading glasses, universal availability of, 183
Reasonable accommodation
  legal definition of, 43
  tax credits and, 44
Recycled clothes, 237
Rehabilitation Act, 7
Rehabilitation Act (1973), Section 502, 37, 42
Rehabilitation Act (1998), Section 508, 7, 37, 51, 88
  assistive technology mandate of, 123
  difference principle and, 240
  functional performance criteria, 52, 53
  impact of, 54
  implications of, 51
  information, documentation, and support, 52
  software applications, 89
  technical standards, 52
  World Wide Web and, 202, 207
Reich, Robert, 8
Reliability engineering techniques
  design of, 149
  reducing common cause variability using, 162, 163
Remote control units, 93
Remote sensor technology, home health care and, 120
"Remove mindset barriers" initiative, 189
Restaurant menus, 107
Retail environment, self-managing, 191

Index

Retail industry, lean production in, 190
RFID tags, *see* Radio frequency identification tags

S

Safety risks, error-proofing and, 198
SBC, *see* Southwestern Bell Communications
Schiavo, Terri, 221
Scissors, affordances of, 98
Section 508 Buy Wizard, 57, 58, 214
Securities and Exchange Commission, 221
Self-organization mechanism, 168–169
Senses, discrimination capacities for, 91
Sensory perceptions, judgments made based on, 90
Sensory processing, 87
Service delivery systems, building flexibility into, 126
Service flexibility, key dimensions of, 127
Shadow diagram pegboard, 168, 170
shopgoodwill.com, 237
Signal
    detection
        expression of, 90
        physiology of, 86
        sensory discrimination and, 86
    judgments, confusion of, 91
    transformation, 86
Signal-to-noise ratio, 85, 89
Signing avatar programs, marketing of, 122
Sign language presentations, 122
Six Sigma
    origins of, 163, 164
    Wal-Mart's use of, 191, 235
Slips
    definition of, 133
    errors resulting from, 138
Small businesses, use of universal design principles in, 196
Smart inventory containers, 137
Social attitudes, 224
Social capital, 234
Social model, 33
    importance of environment in, 35
    view of disability in, 33, 43
Social and political perspective, universal and accessible design from, 231–244
    evolution of universal design, 232–234
    examples of linkage complexity and changing role of designers, 235–237
    linkages, 231–232

universal and accessible design in relation to society and politics, 237–241
universal design from systems perspective, 234–235
Societal values, 229
Software
    applications, Rehabilitation Act, Section 508, 89
    crashes, 159
    development programs, dropdown help features of, 140
Sony Electronics, 57
Southwestern Bell Communications (SBC), 9, 228
Special causes, 157
Special education training workshop, 43
Speedee Service System, fast food, 191
Stable and predictable design, 157–177
    design strategies, 159–170
        reduce common cause variability associated with person's interaction with product or process, 161–162
        reduce common cause variability associated with process use, 163–170
        reduce common cause variability using quality control and reliability engineering techniques to ensure proper functioning of product, 162–163
        work to establish national and international standards for products, processes, and services to reduce common cause variability, 159–161
    discussion, 157–159
    examples of strategies and synergies, 170–174
        airport, 172–174
        meter reading, 170–172
    principle, 157
Standardized operating procedures, 169
Stevens' power law, 87
    categorization of stimuli, 88
    sensory processing and, 87
    signal detection expressed by, 86, 90
Sticky Keys, 124
Subaru vehicles, manufacturing site for, 188
Substitutive representation, color as, 100
Sun Microsystems, 57
Supply chain
    businesses, 192
    logistics, 235
"Sustainable and light" systems, 234

Symbols
  informational value of, 102
  universally understood, 106
System(s)
  design, agile, 153–154
  model, view of disability in, 34
  "sustainable and light", 234

T

Tactile feedback, 83
Tactile roughness, 87, 88
Talking scale system, 9
Tash Inc., 57
Task
  analysis tools, NVAA reduction and, 153
  complexity, cognitive demands and, 95
  sequencing system (TSS), 138
  structure, 110
Taste, 88
Tax credits, employer receiving, 44
TDDs, *see* Telecommunications Device for the Deaf
Technology Transfer, 183
Telecommunication(s)
  accessibility, lack of, 228
  equipment, compatibility of, 49
  products, mandates on design of, 51
  products and services, designing of, 47
  relay services (TRS), 46
Telecommunications Act (1996), 37, 41
Telecommunications Act (1996), Section 255, 7, 18, 21, 46, 51, 81
  assistive technology mandate of, 123
  cognitive accessibility, 95
  difference principle and, 240
  examples of multisensory communication options, 84
  input, control, and mechanical functions, 48–49
  product design, development and evaluation, 47–48
Telecommunications Device for the Deaf (TDDs), 46
Telephone(s)
  system, cognitive demands associated with use of, 96
  volume controls attached to, 20
Teletypewriter (TTY), 46, 96, 122
  connectability, 50
  signal compatibility, 50
Tennis elbow, 21
Texas Instruments calculator TI-84, 113
Text-to-Braille systems, 30, 225

Text-to-voice programs, 30, 122
Third Age Suit, Ford Motor Company's, 7, 78
Thought process(es)
  design as, 17
  elements of, 16
  law mandating, 48
  Web designer, 215
Tight tasks, 93
Tool storage, pegboard for, 168
Total quality management (TQM), 166, 167
"Touch-and-go" process, phone service system, 152
Toyota
  commitment to universal design, 4
  production system (TPS), 166
  TOYOTA Universal Design Showcase, 228
  Universal Design Showcase, 4, 5
TPS, *see* Toyota production system
TQM, *see* Total quality management
Trace Research & Development Center, 215
Traffic
  symbols, universally understood, 102
  volume, common cause variation and, 157
TRS, *see* Telecommunications relay services
Truck(s)
  designers, design for women driving trucks, 80
  environmental adjustability in, 123
TRW Systems, 57
TSS, *see* Task sequencing system
TTY, *see* Teletypewriter

U

UFAS, *see* Uniform Federal Accessibility Standards
UL, *see* Underwriters Laboratory, Inc.
Underdeveloped country, example of universal design in, 22
Underwriters Laboratory, Inc. (UL), 57, 162
Uniform Federal Accessibility Standards (UFAS), 39, 42
Union of the Physically Impaired against Segregation (UPIAS), 33
United Nations, Standard Rules, 58
United States of Wal-Mart, 190
Universal design, *see also* Hierarchical structure of universal design principles
  accessible design vs., 6
  approach, airport terminals using, 173
  CAST, 127
  cost-effectiveness of, 8

Council of Europe, Committee of Ministers, definition of, 227
definition of, 5, 17, 226–227
equitable, 180
ethical dimension of, 232
example of in underdeveloped country, 22
force driving use of, 9
OXO principles of, 8
politics and, 237
presence of in global market, 22
principle(s), 67–68
　application in small businesses, 196
　ecological concerns and, 233
　efficiency as, 154
　fundamental premise of, 232
　global initiatives directed at teaching, 58
　hierarchical structure of, 68
　process-oriented, 143
　Web Accessibility Initiative and, 206–207
society's conceptualization of, 229
Southwestern Bell Communication commitment to, 9
strategies, fundamental premise of, 67
Toyota's commitment to, 4
workplace, 75
Universal design/accessible design/adaptable design, 15–25
　accessible design, 18
　adaptable design, 18–19
　definition of design, 15–17
　examples of entities from each design category, 20–21
　　accessible design, 20
　　accessible design and universal design, 20
　　adaptable, accessible, and universal design, 21
　　adaptable design, 21
　　adaptable design and accessible design, 20–21
　　universal design, 20
　relationship between general design, universal design, and accessible design, 19–20
　universal design in global market, 22–23
UPIAS, *see* Union of the Physically Impaired against Segregation
Usability, attributes of, 17
Usability.gov, 214
UsableNet, Inc., 214
U.S. Access Board, 11, 57, *see also* Access Board

U.S. Food and Drug Administration (FDA), 162

V

Van conversions, wheelchair accessibility provided by, 19, 21
VCR remote controls, 92
Venn diagram, 19
Virtual resource center, 15
VISIO®, 115
Visual controls, common cause variability reduced using, 169
Visual information, availability of, 84
Visualization tools, 115
Visual learners, 115
Visual prompts, 84
Voice
　prompting, 9
　telephones, 46

W

WAI, *see* Web Accessibility Initiative
Waiting times, reduction of, 147
Wal-Mart, 237
　competition with, 96
　early stores, 190
　inventory management system of, 191
　purchasing power of, 191
Warning strategies, sensing element in, 137
Watchfire Corporation, Bobby, 214
W3C, *see* World Wide Web Consortium
WCAG, *see* Web Content Accessibility Guidelines
Web, *see also* World Wide Web
　accessibility, ADA and, 46
　-based material, examples of on accessible design, 11
　-based services, marketing of signing avatar programs for, 122
　-based tutorials, 54
　designers, thought process of, 215
　page from hell, 108, 109
Web Accessibility Initiative (WAI), 202, 203, 227
　accessibility guidelines, 202, 204
　conformance levels, 205
　guidelines and universal design principles, 206–207
Web Content Accessibility Guidelines (WCAG), 205, 209
Weber's law, 86, 87, 90
Wheelchair accessibility

bathrooms with, 124
van conversions for, 19, 21
WHO-ICF, see World Health Organization International Classification of Functioning and Disability
Wind-up radio, 22
Wireless phone manufacturers, 50
Women truck drivers, 80
Word processing program(s)
   accessibility to, 144
   error made using, 138
   learning to use, 142, 143
   selection of, 141
   spellcheck features of, 152
Work
   global distribution of, 198
   processes, flexibility and, 125
   standardized, 169
   steps, product value and, 149
Worker safety, OSHA requirements and, 198
Workforce Investment Act (1998), 18
Workplace
   assistive technology in, 31
   communication of information in, 85
   entities, 188
   ergonomics and, 75
   example of universal design in, 9
   general principle of universal design in, 75
   importance of body size in, 79
   morale, 132
   organization
      chef's use of, 169
      lean production and, 168
      McDonald's use of, 192
   OSHA requirements for, 79
Workplace, universal and accessible design in, 187–199
   consumer applications versus workplace applications, 187–188
   design tools for work, 197
   examples, 188–196
      Fuji Heavy Industries Ltd., 188–190
      retail and fast food industries, 190–192
      small company, 192–193
      small retail operations, 193–196
   nature of industry, size of company, and scale of operations, 196–197
World Health Organization International Classification of Functioning and Disability (WHO-ICF), 30, 35
   major dimensions of, 30, 31
   model, 54, 216, 226
   statement on relationship between disability and environmental factors, 31
World Wide Web, 201–217, see also Web
   accessibility guidelines, 204, 206–207
   additional resources, 215
   authoring tools, 213–214
   content accessibility guidelines.
   decentralization, 204
   design community, procedures of, 201
   design principles, 203–204
   emerging products, standards, guidelines, and Web design recommendations, 211–213
   environment, creation of stable, 205
   grid services, 211
   infrastructure, 202
   inventor of, 203
   Section 508 requirements, 207–211
   United States government resources, 214
World Wide Web Consortium (W3C), 46, 201–202
   Authoring Tool Accessibility Guidelines 1.0, 213
   goal of, 202
   principle tasks of, 203
   recommendations, 203
   Web Accessibility Initiative, 11, 202, 203, 227
Writing process, 141, 142
WYSIWYG editing tool, 213

X

Xerox Corporation, 57